THE FRONTIERS OF
HUMAN BRAIN RESEARCH

HEARING

BEFORE THE

SUBCOMMITTEE ON RESEARCH AND TECHNOLOGY

COMMITTEE ON SCIENCE, SPACE, AND
TECHNOLOGY

HOUSE OF REPRESENTATIVES

ONE HUNDRED THIRTEENTH CONGRESS

FIRST SESSION

———

WEDNESDAY, JULY 31, 2013

———

Serial No. 113–45

———

Printed for the use of the Committee on Science, Space, and Technology

Available via the World Wide Web: http://science.house.gov

———

U.S. GOVERNMENT PRINTING OFFICE

82–224PDF   WASHINGTON : 2013

COMMITTEE ON SCIENCE, SPACE, AND TECHNOLOGY

HON. LAMAR S. SMITH, Texas, *Chair*

DANA ROHRABACHER, California
RALPH M. HALL, Texas
F. JAMES SENSENBRENNER, JR.,
  Wisconsin
FRANK D. LUCAS, Oklahoma
RANDY NEUGEBAUER, Texas
MICHAEL T. McCAUL, Texas
PAUL C. BROUN, Georgia
STEVEN M. PALAZZO, Mississippi
MO BROOKS, Alabama
RANDY HULTGREN, Illinois
LARRY BUCSHON, Indiana
STEVE STOCKMAN, Texas
BILL POSEY, Florida
CYNTHIA LUMMIS, Wyoming
DAVID SCHWEIKERT, Arizona
THOMAS MASSIE, Kentucky
KEVIN CRAMER, North Dakota
JIM BRIDENSTINE, Oklahoma
RANDY WEBER, Texas
CHRIS STEWART, Utah
VACANCY

EDDIE BERNICE JOHNSON, Texas
ZOE LOFGREN, California
DANIEL LIPINSKI, Illinois
DONNA F. EDWARDS, Maryland
FREDERICA S. WILSON, Florida
SUZANNE BONAMICI, Oregon
ERIC SWALWELL, California
DAN MAFFEI, New York
ALAN GRAYSON, Florida
JOSEPH KENNEDY III, Massachusetts
SCOTT PETERS, California
DEREK KILMER, Washington
AMI BERA, California
ELIZABETH ESTY, Connecticut
MARC VEASEY, Texas
JULIA BROWNLEY, California
MARK TAKANO, California
ROBIN KELLY, Illinois

————

SUBCOMMITTEE ON RESEARCH AND TECHNOLOGY

HON. LARRY BUCSHON, Indiana, *Chair*

STEVEN M. PALAZZO, Mississippi
MO BROOKS, Alabama
RANDY HULTGREN, Illinois
STEVE STOCKMAN, Texas
CYNTHIA LUMMIS, Wyoming
DAVID SCHWEIKERT, Arizona
THOMAS MASSIE, Kentucky
JIM BRIDENSTINE, Oklahoma
LAMAR S. SMITH, Texas

DANIEL LIPINSKI, Illinois
FEDERICA WILSON, Florida
ZOE LOFGREN, California
SCOTT PETERS, California
AMI BERA, California
DEREK KILMER, Washington
ELIZABETH ESTY, Connecticut
ROBIN KELLY, Illinois
EDDIE BERNICE JOHNSON, Texas

CONTENTS

Wednesday, July 31, 2013

IV

Appendix I: Additional Material for the Record

THE FRONTIERS OF HUMAN BRAIN RESEARCH

WEDNESDAY, JULY 31, 2013

House of Representatives,
Subcommittee on Research and Technology
Committee on Science, Space, and Technology,
Washington, D.C.

The Subcommittee met, pursuant to call, at 11:06 a.m., in Room 2318 of the Rayburn House Office Building, Hon. Larry Bucshon [Chairman of the Subcommittee] presiding.

LAMAR S. SMITH, Texas
CHAIRMAN

EDDIE BERNICE JOHNSON, Texas
RANKING MEMBER

Congress of the United States
House of Representatives

COMMITTEE ON SCIENCE, SPACE, AND TECHNOLOGY

2321 RAYBURN HOUSE OFFICE BUILDING

WASHINGTON, DC 20515-6301

(202) 225-6371
www.science.house.gov

Subcommittee on Research and Technology

The Frontiers of Human Brain Research

Wednesday, July 31, 2013
11:00 a.m. to 1:00 p.m.
2318 Rayburn House Office Building

Witnesses

Dr. Story Landis, *Director of National Institute of Neurological Disorders and Stroke, National Institutes of Health*

Mr. Michael McLoughlin, *Deputy Business Area Executive, Research and Exploratory Development at Applied Physics Laboratory, Johns Hopkins University, and* **U.S. Air Force Master Sergeant Joseph Deslauriers Jr.**

Dr. Marcus Raichle, *Professor of Radiology, Neurology, Neurobiology and Biomedical Engineering, Washington University*

Dr. Gene Robinson, *Director, Institute for Genomic Biology, Swanlund Chair, Center for Advanced Study Professor in Entomology and Neuroscience, University of Illinois, Urbana-Champaign*

3

U.S. HOUSE OF REPRESENTATIVES
COMMITTEE ON SCIENCE, SPACE, AND TECHNOLOGY
SUBCOMMITTEE ON RESEARCH

HEARING CHARTER

The Frontiers of Human Brain Research

Wednesday, July 31, 2013
11:00 a.m. - 12:30 p.m.
2318 Rayburn House Office Building

1. Purpose

On Wednesday, July 31, 2013, the Subcommittee on Research and Technology will hold a hearing to understand the frontiers and challenges of brain science research, including its potential and limitations for curing brain diseases and rehabilitating those with brain-related injuries and disorders. The hearing will also aim to understand any policy implications from this research, including any implications for the America COMPETES reauthorization.

2. Witnesses

Dr. Story Landis, Director, National Institute for Neurological Disorders and Stroke (NINDS) at the National Institutes of Health (NIH)

Michael McLouglin, Deputy Business Area Executive, Research and Exploratory Development, Applied Physics Laboratory at Johns Hopkins University and **U.S. Air Force Master Sergeant Joseph Deslauriers Jr.**

Dr. Marcus Raichle, Professor of Radiology, Neurology, Neurobiology and Biomedical Engineering, Washington University in St Louis

Dr. Gene Robinson, Professor in Entomology and Neuroscience and Director of the Institute for Genomic Biology, University of Illinois at Urbana-Champaign

3. Hearing Overview

Understanding the human brain remains one of the most complex tasks facing the medical sciences community. Throughout the 19th and 20th century, much progress was made by breaking the brain down into various components, with individual neurons viewed as the fundamental unit for human brain activity. The number of these neurons is roughly a hundred billion; the number of contacts between neurons is a hundred trillion. The brain is a complex organ that processes and receives electrical, chemical and mechanical inputs and outputs. The average neuron receives thousands of distinct inputs, with each neuron connecting to many other neurons; however the exact physical "circuitry" of these individual components is unknown.

Taxpayer funded research in neuroscience, the scientific field that studies the nervous system, has been crucial to advancing our understanding into the workings of the brain. During

FY 2013, the National Institutes of Health (NIH) spent over $5.6 billion in supporting neuroscience-related research. Brain science is an inter-disciplinary field, with important contributions from fields as diverse as electrophysiology, imaging, molecular biology, biochemistry, physics and applied mathematics. Each of these disciplines has enriched our understanding of the brain, thereby allowing researchers to move towards an integrated picture of the brain's behavior through translational research and medicine.

Attempts to map the brain into distinct areas, each with its specific function, is a scientific approach that has existed for over a century. In the mid-1800s, brain science focused on discovering and "mapping" the functions of the cerebral cortex using a variety of methods and techniques that were available at that time. This field is called phrenology, and the mapping paradigm of localizing cerebral functions within the brain was its primary focus. Today, one of the main challenges is moving from a static to a more dynamic view of the brain. Other challenges include finding accurate and repeatable medical tests for diagnosing brain disorders.[1]

The experimental tools available to the biological sciences are rapidly growing, with many in the brain research community viewing them as the key to unlocking the next breakthrough. This evolving experimental and computational toolkit, much of it funded by federal science agencies, is addressed at understanding the brain at various time and length scales. This includes the Human Connectome Project, a $30M NIH-funded endeavor to map long-distance neural pathways in the brains of 1,200 healthy adult humans using functional magnetic resonance imaging (fMRI). Other advances include two-photon functional imaging and next generation brain-machine interface using integrated neurophotonics and nanoparticles.

The Defense Advanced Research Projects Agency (DARPA) has recently been funding research projects that aim to diagnose and develop therapeutic responses for brain and spinal cord injury. In particular, DARPA has sponsored the development of robotic arms that are allowing returning veterans, who have lost limbs, to lead functioning lives. For example, sensors on the skin can detect the brain's signals from the nerves and use those signals to control a robotic arm, without an invasive surgical procedure. Other neuro-prosthetic technologies include implanting sensors inside the human brain to control remote robotic appendages with only a person's thoughts.

Economic Impact

The economic implications for this research are significant. The monetary cost of dementia, including Alzheimer's disease, in the United States ranges from $157 billion to $215 billion annually. Dementia is more costly to the nation than either heart disease or cancer.[2] As our nation's population grows older, the costs of dementia related diseases could double by 2040. In a 2008, Dr. Thomas Insel, the director of the National Institute for Mental Health, estimated that the total cost of serious mental illness in the US exceeds $317 billion per year.[3] These annualized costs include: an estimated $193 billion in lost earnings, roughly $100 billion in direct health care costs, and another $24 billion in disability benefits. This estimate ignores other major factors that increase the cost burden to our nation, including: homelessness, incarceration, substance abuse and other addictions, all of which are associated with mental illness. The US

---

[1] http://www.npr.org/blogs/health/2013/07/08/198086616/MENTAL-ILLNESS-BIOMARKERS
[2] http://www.rand.org/news/press/2013/04/03.html
[3] http://ajp.psychiatryonline.org/article.aspx?articleid=99862

National Bureau of Economic Research has estimated that 38 percent of all alcohol, 44 percent of all cocaine, and 40 percent of all cigarettes are consumed by people with a mental illness.[4]

BRAIN Initiative

In April, the Obama Administration announced the Brain Research through Advancing Innovative Neurotechnologies (BRAIN) Initiative prior to presenting his FY 2014 budget request; this is a joint research program between federal science agencies— NSF, NIH, and DARPA —and private sector partners in research of brain disorders ranging from Alzheimer's and Parkinson's disease, epilepsy, autism, and injuries. The initiative has been met with some skepticism by the research community for its intent and how it might divert funds from other research.[5] While the NSF's FY 2014 budget presentation to the Committee did not highlight the Foundation's contribution to the Administration's BRAIN Initiative as a priority, $14 million in additional spending is requested for Cognitive Science and Neuroscience. The NIH and Defense Advanced Research Projects Agency (DARPA) will each receive $40 million and $50 million in funding respectively. Private organizations are also involved with this program, including $60 million from the Allen Institute for Brain Science, $30 million from the Howard Hughes Medical Institute, $28 million from the Salk Institute for Biological Studies, and $4 million from the Kavli Foundation.

Supporters of the BRAIN Initiative have compared the project to the Human Genome Project, whose main benefit has been the development of technologies to quickly and accurately screen the genomes of individual clients. However, the actual benefit from the Human Genome Project has been limited so far in that no diseases have yet to be cured as a result of the human genome map. Critics of the BRAIN Initiative have mentioned that this investment needs more specific goals or end-points. The central question is not about the technology, but whether what the BRAIN initiative's goal of "mapping the human brain" provides an accurate reflection of how the brain works, and whether brain-mapping is a valid or outmoded paradigm to make progress in the field. Further, advocates for the BRAIN Initiative need to communicate the long-term nature of such promising research and when they can reasonably expect certain breakthroughs, especially given the level of public investment.

European Commission's Human Brain Project

There is world-wide interest in understanding the brain. The Human Brain Project has been officially selected as one of the European Commission's two *Future and Emerging Technologies Flagship Projects*. The goal of the EU's Human Brain Project is to consolidate existing knowledge about the human brain and to reconstruct the brain using a combination of modeling and computational simulations; the project will rely heavily on the use of supercomputers. The models offer the prospect of a new understanding of the human brain and its diseases and of completely new computing and robotic technologies. The Human Brain Project is planned to last ten years (2013-2023).

---

[4] http://www.nber.org/digest/apr02/w8699.html
[5] http://www.npr.org/2013/04/05/176303594/researchers-question-obamas-motives-for-brain-initiative

Issues for Consideration

The hearing will examine the latest developments in the area of brain science research, and stress the interdisciplinary approach that is necessary for understanding the complexities of the human brain. Witnesses have been invited to give demonstrations of their technologies and discuss the future of brain science research. They will also comment on current and future federal initiatives in the area of brain science research.

Chairman BUCSHON. The Subcommittee on Research and Technology will come to order.

Good morning. Welcome to today's hearing entitled "The Frontiers of Human Brain Research." In front of you are packets containing the written testimony, biographies, and truth-in-testimony disclosures of today's witnesses.

I now recognize myself for an opening statement.

I would like to welcome everyone to today's Research and Technology Subcommittee hearing on the frontiers of human brain research. As a doctor, I know firsthand there are many complexities surrounding the human body and understanding the human brain is one of the most challenging problems facing the scientific and medical communities. This problem will likely require an interdisciplinary and multifaceted approach with the right scientific questions being asked and debated and clear goals and endpoints being articulated. The creative drive of American science is the individual investigator, and I have faith they will continue to tackle, understand, and contribute original approaches to these problems.

We are hopeful that brain research will have important policy implications. Brain disorders such as Alzheimer's, Parkinson's, autism, epilepsy, dementia, stroke, and traumatic brain injury have an enormous economic and personal impact for the affected Americans. For example, Alzheimer's disease, a severe form of dementia and the sixth leading cause of death in the United States, affects the 5.1 million Americans that have the disease along with their friends and family who watch their loved one suffer from its effects. And my best friend from high school's grandmother was one of those people.

I want to stress the personal effect of this research, which to me is much more important as a medical doctor but cannot be easily quantified. During my visits to Walter Reed Medical Center and subsequently Bethesda after Walter Reed closed, I have met with many brave young men and women who unfortunately have suffered traumatic brain injury as well as lost limbs because of their service to our country in Iraq and Afghanistan. Technologies, like the ones we will hear about today, will allow these young men and women to transition to the workplace, enabling these individuals to lead productive, independent, and fulfilling lives. This is why I think it is so important to continue to support research.

I want to stress my support for brain science research, in particular understanding neurological disorders and diseases from an interdisciplinary perspective. As our witnesses will testify today, brain science has benefited enormously from fields as diverse as applied mathematics, computer science, physics, engineering, molecular biology, and chemistry. More importantly, basic science research results from NSF-funded research will be the future experimental tools for hypothesis-based, data-driven research for brain science researchers.

I see this as an important opportunity for continuing interdisciplinary work between the various Federal science agencies, including NSF, NIH, and DARPA and I hope to see more collaboration and productive research opportunities.

Our witnesses today reflect the wide spectrum of research in brain science and the richness in this field. I would like to thank the witnesses for being here today and taking the time to offer their perspectives on this important topic. At this point, I also would like to thank Ranking Member Lipinski and everyone else for participating in today's hearing.

And I will now recognize Ranking Member Lipinski for his opening statement.

[The prepared statement of Mr. Bucshon follows:]

PREPARED STATEMENT OF SUBCOMMITTEE ON RESEARCH AND TECHNOLOGY CHAIRMAN LARRY BUCSHON

I would like to welcome everyone to today's Research and Technology Subcommittee hearing on the frontiers of human brain research.

As a doctor, I know firsthand there are many complexities surrounding the human body and understanding the human brain is one of the most challenging problems facing the scientific and medical communities. This problem will likely require an inter-disciplinary and multifaceted approach with the right scientific questions being asked and debated and clear goals and endpoints being articulated. The creative drive of American science is the individual investigator, and I have faith they will continue to tackle, understand and contribute original approaches to these problems.

We are hopeful that brain research will have important policy implications. Brain disorders such as Alzheimer's, Parkinson's, autism, epilepsy, dementia, stroke, and traumatic brain injury have an enormous economic and personal impact for affected Americans.

For example, Alzheimer's disease—a severe form of dementia and the sixth leading cause of death in the US—affects the 5.1 million Americans that have the disease along with their friends and family who watch their loved one suffer from its effects. The average annual cost of care for people with dementia over 70 in the US was roughly between $157 and $210 billion dollars in 2010.

More importantly, I want to stress the personal effect of this research, which to me is much more important as a medical doctor, but cannot be easily quantified. During my visits to Walter Reed Medical Hospital, I have met many brave young men and women who have unfortunately lost their arms and legs in Iraq and Afghanistan. Technologies, like the ones we will hear about today, will allow these young men and women to transition to the workplace, enabling these individuals to lead productive, independent, and fulfilling lives. This is why I think it's so important to continue supporting this research.

I want to stress my support for brain science research, in particular understanding neurological disorders and diseases from an interdisciplinary perspective. As our witnesses will testify today, brain science has benefited enormously from fields as diverse as applied mathematics, computer science, physics, engineering, molecular biology, and chemistry. More importantly, basic science research results from NSF funded research will be the future experimental tools for hypothesis-based data-driven research for brain science researchers.

I see this as an important opportunity for continuing interdisciplinary work between the various federal science agencies, including the NSF, NIH and DARPA and I hope to see more collaboration and productive research opportunities

Our witnesses today reflect the wide spectrum of research in brain science and richness in this field. I'd like to thank the witnesses for being here today and taking time to offer their perspectives on this important topic. I'd also like to thank Ranking Member Lipinski and everyone else participating in today's hearing.Before I conclude today's hearing, I would like to recognize and thank Melia Jones. I appreciate your work on this Subcommittee for the last 2 years, and wish you all the best in your future endeavors. We hate to lose you, but Texas will gain a good friend.

Mr. LIPINSKI. Thank you, Chairman Bucshon, for holding this hearing and to all the witnesses for being here today. And I thank you for your flexibility in moving this hearing back an hour.

I don't think there is anyone in this room who hasn't marveled at the complexity of the human brain. I know opening up with that sentence lends itself to a lot of jokes about Congress, so you can insert your own joke here, but what we are really concerned about are brain diseases especially that befall so many people. And we all know it may one day wreak havoc on our own lives, in addition to that, obviously other brain injuries that occur. And especially as lawmakers, we are responsible for making sure our returning servicemen and women are taken care of after they have so bravely risked their own lives, especially we worry about the thousands of returning from Iraq and Afghanistan and previous conflicts with traumatic brain injury and long-term mental distress.

In April of this year, President Obama announced the BRAIN Initiative, an interagency collaboration between DARPA, NIH, and NSF to accelerate what we know about human brain function and its connection to behavior. Each of these agencies has important research activities that it can bring to the table. The NSF, for example, will help further research developing probes on a molecular scale that can map the activity of neural networks. They can also bring computer scientists to the task as well to help understand the functions of the estimated 100 billion neurons and 100 trillion connections within the human brain.

As we take a broad look at Federal support for neuroscience research in general and the BRAIN Initiative in particular, I believe it is valuable for the Members of this Committee to hear from experts who can speak to the roles of all key agencies, including DARPA and NIH. Three of the witnesses are highly qualified to speak to NIH's role. Mr. McLoughlin has long been funded by DARPA.

However, the only BRAIN Initiative agency wholly within this Committee's jurisdiction is the National Science Foundation. It is unfortunate that the NSF was not invited to participate on today's panel, but I am especially grateful to Dr. Robinson for being here today to help us better understand NSF's unique and important role in supporting neuroscience research. And I know that Chairman Bucshon had duly noted the important role of NSF in his opening statement.

The idea of connecting what is happening in our brain at the molecular level with how we feel, think, and remember and act is known as integrating across scales. We can bring to the neuroscience table all the smart computer scientists, engineers, and mathematicians we can find, and we do need them, but if we don't also have the behavioral experts there to validate brain function models with what we know about actual human behavior, those models might not be worth the laptops they are written on.

As the one agency that funds basic research in all fields of science and engineering, including the social and behavioral sciences, integrating across scales is one of the strengths that NSF brings to the BRAIN Initiative.

While none of the witnesses were asked to address educational needs and opportunities in neuroscience, this is also an area in which NSF leads the way. And I have some questions related to STEM Ed, and I suspect some of my colleagues will as well.

Thank you again to Chairman Bucshon for holding this hearing and I look forward to the testimony and the discussion.

[The prepared statement of Mr. Lipinski follows:]

PREPARED STATEMENT OF SUBCOMMITTEE ON RESEARCH AND TECHNOLOGY
RANKING MINORITY MEMBER DANIEL LIPINSKI

Thank you Chairman Bucshon for holding this hearing and to all of the witnesses for being here.

I don't think there's anybody in this room who hasn't marveled at the complexity of the human brain. With that wonder also comes worry about the brain diseases that befall so many people, and that we all know could someday wreak havoc on our own lives. And as lawmakers responsible for making sure our returning servicemen and women are taken care of after they have bravely risked their own lives, we worry about the thousands who have returned from Iraq, Afghanistan, and previous conflicts with traumatic brain injury and long-term mental distress.

In April of this year, President Obama announced the BRAIN Initiative, an interagency collaboration between DARPA, NIH, and NSF to accelerate what we know about human brain function and its connection to behavior. Each of these agencies has important research activities that it can bring to the table. The NSF, for example, will help further research developing probes on a molecular scale that can map the activity of neural networks. They can also bring computer scientists to the task as well, to help understand the functions of the estimated 100 billion neurons and 100 trillion connections within the human brain.

As we take a broad look at federal support for neuroscience research in general, and the BRAIN Initiative in particular, I believe that it is valuable for the Members of this Committee to hear from experts who can speak to the roles of all key agencies, including DARPA and NIH. Three of the witnesses are highly qualified to speak to NIH's role, and Mr. McLoughlin has long been funded by DARPA. However, the only BRAIN Initiative agency wholly within this Committee's jurisdiction is the National Science Foundation. It is unfortunate that NSF was not invited to participate on today's panel, but I am especially grateful to Dr. Robinson for being here to help us better understand NSF's unique and important role in supporting neuroscience research.

The idea of connecting what's happening in our brain at the molecular level with how we feel, think, remember, and act is known as "integrating across scales." We can bring to the neuroscience table all of the smart computer scientists, engineers, and mathematicians we can find. And we do need them. But if we don't also have the behavioral experts there to validate brain function models with what we know about actual human behavior, those models might not be worth the laptops they're written on.

As the one agency that funds basic research in all fields of science and engineering, including the social and behavioral sciences, integrating across scales is one of the strengths that NSF brings to the BRAIN Initiative. While none of the witnesses were asked to address educational needs and opportunities in neuroscience, this is also an area in which NSF leads the way. I have questions related to STEM education and I suspect some of my colleagues will as well.

Thank you again Chairman Bucshon for holding this hearing and I look forward to the testimony and discussion.

Chairman BUCSHON. Thank you.

I now recognize Mr. Stockman.

Mr. STOCKMAN. I just want to thank the Chairman, Mr. Bucshon, for doing this. And as I mentioned to you earlier, I took care of my father for eight years who had Alzheimer's, and, as you know, some say that disease is hereditary, so hurry up and do your work.

And the other thing is that I was listening to National Public Radio which commented on the President's Initiative, and I hope that it is more than just window dressing that we have here and that we have real research. I appreciate you coming out today and I really appreciate the Ranking Member and the Chairman for having this hearing. I yield back. Thank you.

Chairman BUCSHON. Thank you. If there are Members who wish to submit additional opening statements, your statements will be added to the record at this point.

At this time I am now going to introduce our witnesses.

Our first witness today is Dr. Story Landis. Since 2003, she has been the Director of the National Institute of Neurological Disorders and Stroke. Prior to her appointment at NINDS for short, she was a Professor and Chairwoman of the Department of Neurosciences at Case Western Reserve University School of Medicine in Cleveland, Ohio. She has made many fundamental contributions to understanding the developmental interactions required for synapse formation. I understand that but many in the room may not. But she is an elected fellow of the American Academy of Arts and Sciences and the Institute of Medicine for the National Academy of Sciences.

Our second witness today is Dr. Michael McLoughlin who is a Deputy Business Area Executive for the Johns Hopkins University Applied Physics Laboratory Research and Exploratory Development—in the exploratory development area. In addition to this position, Mr. McLoughlin teaches both program management and systems engineering at Johns Hopkins University Whiting School of Engineering. In 2009 he assumed leadership responsibilities for DARPA's revolutionizing prosthetics program and is leading efforts to transition use of these technologies to human subjects. Mr. McLoughlin is a graduate of the University of Delaware where he received both his bachelor's and master's degrees.

Also with him is Air Force Master Sergeant Joseph Deslauriers, an Explosive Ordnance Disposal Technician who also will be giving a short testimony on how some of these technologies have impacted the quality of his own life. He earned the Silver Star for Gallantry in Action while serving in Afghanistan on September 23, 2011.

Our third witness is Professor Marcus Raichle, who is currently the Professor of Radiology, Neurology, Neurobiology and Biomedical Engineering at Washington University in St. Louis. Professor Raichle has led world-class efforts to define the frontiers of cognitive neuroscience through the development and use of functional brain imaging techniques. He has also pioneered the concept of the default mode of brain function and has invigorated studies of intrinsic functional activity. Professor Raichle is a member of the U.S. Academy of Science, the American Academy of Arts and Sciences, and the Institute of Medicine.

And our final witness is Professor Gene Robinson, who received his doctorate degree from Cornell in 1986, and since 1989 has been on the faculty of the University of Illinois in Urbana-Champaign where he is the University Swanlund Chair and the Director for Genomic Biology. He has pioneered the application of genomics to the study of behavior. He is the author or co-author of over 250 publications. Professor Robinson is a member of the U.S. National Academy of Science and the American Academy of Arts and Sciences. In addition, he received the National Institute's Pioneer Award.

Thanks again for all of our witnesses for being here this afternoon. It is a very distinguished panel. I am looking forward to your testimony.

As our witnesses should know, spoken testimony is limited to five minutes after which the Members of the Committee will have five minutes each to ask questions.

I now recognize Dr. Landis for five minutes to present her testimony.

TESTIMONY OF DR. STORY LANDIS, DIRECTOR OF NATIONAL INSTITUTE OF NEUROLOGICAL DISORDERS AND STROKE, NATIONAL INSTITUTES OF HEALTH

Dr. LANDIS. Good morning, Chairman Bucshon, Ranking Member Lipinski, and embers of the Subcommittee. I want to thank you very much for your opportunity to provide testimony today on the frontiers of human brain research. This is an incredibly exciting area of research with profound implications for our basic understanding of the brain and also for treating brain disorders.

So as you have heard, many people regard understanding how the human brain works as the last great frontier in biological and biomedical sciences. The brain is an extraordinary organ that allows us to see, hear, reason, remember. The best estimates are that these functions and many others are performed by somewhere between 80 and 100—100 billion nerve cells that are connected with each other, each nerve cell, neuron, making more than 1,000 connections with other neurons.

Now, it is not just chaos in the brain. These neurons are organized in neural circuits. You could almost think of them as living modifiable circuit boards which process and integrate different kinds of information to control behavior, mental and physical. And in fact, if you think about the brain, basically the brain is the organ that controls all kinds of behavior.

In the past decade we have made extraordinary advances in developing tools to visualize brain circuits and to dissect their function. One of these tools is diffusion magnetic resonance imaging, and this reveals medium to long-range connections between brain regions and therefore provides a wiring diagram of the human brain. And NIH is currently funding the human brain Connectome Project to create a publicly available database of wiring diagrams for 1,200 people, which will serve as a resource for scientists throughout the world. If I could have the slide please. Can you make it rotate?

[Slide.]

13

MISSING

This is one piece of the human Connectome that was obtained as part of the Connectome Project. Each of those different-colored fibers reflects a different set of connections. This is only a subset of the connections and it is focused primarily on the connections that actually wire together different parts of the cortex. In other studies, we have learned how to actually manipulate the function of neurons, specific populations of neurons and circuits and to define their particular roles.

Now, neuroscience, the study of the brain, has from its very earliest origins been multidisciplinary. Neuroanatomy and neurophysiology and creating that image that you just saw required physicists, engineers, mathematicians, statisticians, as well as a neuroscientists. And just as the science is multidisciplinary, support for brain science is provided by multiple agencies as appropriate for their mission.

So consistent with the NIH's mission to seek fundamental knowledge about the nature and behavior of living systems and the application of that knowledge to enhance health, lengthen life, and reduce illness, NIH funds brain research from the very most basic like ion channels and how neurons get generated during development, how you turn stem cells into neurons to Phase III clinical trials.

Now, my Institute, NINDS, funds research on a large number of neurological disorders, including amyotrophic lateral sclerosis—Lou Gehrig's disease—Parkinson's disease, and Alzheimer's disease. These are inexorably progressive disorders that take away our ability to move, reason, and remember. And we also fund research on a host of rare diseases. We are making progress. Stroke prevention and treatment reduced death from stroke by 40 percent between 1999 and 2009. We have treatments for multiple sclerosis that actually slow progression. We have symptomatic treatments for Parkinson's and many effective drugs that stop seizures.

The NINDS works closely with many other NIH institutes to ensure that we are an aggregate making the best possible investment in brain sciences. There are also strong and effective collaborations between NIH and other agencies. Nine NIH institutes and seven NSF directorates support an innovative grant program, collaborative research, and computational neuroscience, and this grant program requires a wet bench experimentalist working with someone who is a theoretician.

So progress in understanding how the human brain works and addressing diseases that affect the brain will require the development of new tools to allow us to get a dynamic picture of how the brain works in real time, how the individual cells and complex neural circuits interact, and how do they do it at the speed of thought? And we simply don't have the tools to know how to do this. That is the goal of the BRAIN Initiative, brain research advances through innovative neurotechnologies.

Thank you very much for your attention.

[The prepared statement of Dr. Landis follows:]

DEPARTMENT OF HEALTH AND HUMAN SERVICES

NATIONAL INSTITUTES OF HEALTH

The Frontiers of Human Brain Research

Testimony before the

U.S. House of Representatives

Committee on Science, Space, and Technology

Subcommittee on Research

Story C. Landis, Ph.D., Director

National Institute of Neurological Disorders and Stroke

July 31, 2013

Mr. Chairman and Members of the Committee:

I am pleased to testify today about advances in neuroscience, the role of the National Institutes of Health (NIH), and opportunities to accelerate progress through the *Brain Research through Advancing Innovative Neurotechnologies (BRAIN) Initiative*, which the President announced as part of his Fiscal Year (FY) 2014 Budget.

THE IMPORTANCE OF NEUROSCIENCE RESEARCH

Diseases of the brain and nervous system impose an enormous burden on individuals, families, and society. The Institute for Health Metrics and Evaluation, an independent research center, estimates that brain disorders (neurologic, substance abuse, and mental and behavioral disorders) are the number one source of disability globally from all medical causes in those aged 15 to 49 years.[1] The costs of dementia alone—by one recent estimate at $159 billion to $215 billion in the United States for 2012—already rival those of cancer and heart disease, and could rise dramatically in coming decades with a growing elderly population.[2]

Although the burden of nervous system diseases is daunting, the complementary efforts of the NIH and the private sector are making real progress. For example, the age adjusted death rate for stroke fell by 70 percent over the last 50 years, and by approximately 37 percent just from 1999 to 2009.[3] NIH research contributed to this progress by identifying risk factors, determining through many clinical trials which preventive measures are most effective for people with specific risk profiles, and developing the only effective emergency treatment for stroke. New treatments are also available or in testing for many other brain diseases, including Alzheimer's disease, Parkinson's disease, and epilepsy. For example, 20 years ago there were no effective drugs for multiple sclerosis, and now there are nine that slow progression of the disease.

---

[1] Global Burden of Disease Compare, http://www.healthmetricsandevaluation.org/gbd/visualizations/country.
[2] New England Journal of Medicine 368:1326, 2013
[3] Circulation 134:e6-245, 2013.

ADVANCES AND OPPORTUNIES IN NEUROSCIENCE

NIH supports neuroscience research to reduce the burden of brain disorders, from basic research to understand how the brain and nervous system work and what goes wrong as a result of disease, through translational and clinical research to develop and test candidate therapies in the laboratory and in human trials. Many NIH Institutes and Centers support neuroscience research, as appropriate to their missions. Basic neuroscience research, which is my focus today, is an especially critical aspect of the NIH mission because the outcomes and applications of basic research are often too far upstream and high-risk to attract substantial private investment. Basic studies lay the foundation for the development of better diagnosis, treatment, and prevention of neurological and mental health disorders by uncovering disease mechanisms and identifying potential targets for intervention. This research is inherently interdisciplinary and spans multiple levels of analysis, from the intact human brain to single cells, genes, and molecules.

Until recently, the living human brain was largely inaccessible to direct study. Neuroscientists inferred how the human brain works from animal studies and from indirect observations of the human brain, including behavioral experiments, electrical recordings of brain activity from outside the skull, and the consequences of disease and injury to particular parts of the brain. Over the last two decades, advances in brain imaging, which build on developments in physics, engineering, mathematics and other disciplines, have revolutionized both clinical care and neuroscience research. Dr. Marcus Raichle, a pioneer in brain imaging, will testify at this hearing about how brain imaging allows increasingly detailed studies of structure, function, connectivity, and even biochemistry in the living human brain.

Basic neuroscience research focused on molecules, genes, and single cells has also made remarkable progress. Studies at the resolution of single atoms reveal how ion channels control the flow of electrical currents in nerve cells and how receptor molecules on cells sense chemical signals. Just last year, a Nobel Prize recognized the importance of research on a broad class of receptors, called G-protein coupled receptors. Many drugs now in use target these receptors and ion channels, and understanding their structures informs better drug design. The discovery of hundreds of gene defects that cause disease has led to faster diagnoses for families confronting rare neurological disorders. Findings from genetics also lead to better understanding of mechanisms of disease and to rational strategies for the development of therapies by NIH and the private sector. In one dramatic example that may be a harbinger of the future, researchers using "next generation" sequencing methods, which rely on advances in computational analysis, decoded the entire genome of twins with a rare form of dystonia, a movement disorder characterized by abnormally sustained muscle contractions. With insight from this analysis, the research team quickly discovered that a known drug dramatically improved the twins' lives.

Understanding how brain circuits function resides in a middle ground between research on the living human brain and studies of molecules and single cells, and is at the core of some of neuroscience's greatest challenges and opportunities. A unique aspect of the brain, compared to other organs, is the importance of precise connections between brain cells that influence one another's functional output. By one estimate, the human brain has more than 80 billion neurons, each of which may form thousands of synapses (functional connections), with other cells.[4] We perceive, think, and act through the computations performed by these networks of cells, and all but the simplest reflex behaviors in mammals require the concerted activity of many thousands of neurons. Our understanding of how synapses between pairs of nerve cells transmit information has advanced greatly, but research on brain circuits has faced challenges because of the difficulty of monitoring activity in many cells at once. The complete wiring diagram of

---

[4] Ann. Rev. Neurosci.1988. 11 :423-53

neuronal connections has been mapped in one organism, a nematode worm with 302 neurons, but anatomical mapping in mammalian brains has been limited to specific pathways and small parts of circuits. It is as if neuroscientists studying circuits have been trying to watch an HDTV show by observing a few pixels at a time rather than seeing the entire picture at once.

In the last few years, however, new techniques for studying brains in laboratory animals have made mapping the structure and function of brain circuits one of the most exciting and promising areas of neuroscience. Historically, neuroscience has applied advances from multiple disciplines, so it is not surprising that these transformative advances arise from a convergence of neurobiology, genetics, optics, computer science, chemistry, engineering, and other fields and are the products of long term investments by the NIH, the National Science Foundation (NSF), and others. Breakthrough microscopy methods, together with fluorescent indicators that sense specific ions or detect voltages, can simultaneously monitor the activity of thousands of individually resolved cells deep in the brains of living experimental animals. In some experiments, researchers have monitored activity in the same cells for weeks at a time as animals explore their environment and learn. A complementary technique, optogenetics, has dramatically improved researchers' ability to control the activity of specific types of neurons in experimental animals. This method uses genetic engineering to install light-sensitive switches into identified nerve cells, enabling researchers to turn cells' electrical activity on or off with precisely timed light pulses and has quickly become an important tool for neuroscientists exploring neural circuits. For example, a decade ago scientists discovered that adult brains, even in 60 year old people, generate new nerve cells. This year optogenetics revealed that these new cells readily adjust the strength of their synapses and lay down new memories in mice. Optogenetics has also advanced understanding of brain circuits in animal models of Parkinson's disease, post-stroke epilepsy, and the generalization of fearful memories, a process relevant to post-traumatic stress disorder.

New techniques for tracing connections between neurons are also providing unprecedented anatomical maps of the brain's architecture. "Brainbow," for example, labels individual nerve cells and fibers with approximately 100 different colors so that researchers can follow the paths of intertwined nerve fibers through the brain. Another technique called CLARITY[5] renders brain tissue completely transparent, allowing scientists to observe complete trajectories of labeled cells and fiber pathways in the intact brain.

BRAIN INITIATIVE

Although these tools represent remarkable advances, they are not yet adequate to capture the behavior of complete brain circuits functioning in real time in the living human brain. To improve our knowledge and understanding of brain functioning, the President announced the BRAIN Initiative as part of his FY 2014 Budget. Through this initiative, researchers will build on emerging insights from multiple disciplines to develop, disseminate, and apply new tools and technologies that will allow scientists to generate a dynamic, real time picture of entire functioning brain circuits. The BRAIN Initiative will involve the coordinated activities of the NIH, NSF, the Defense Advanced Research Projects Agency (DARPA), the Office of Science and Technology Policy (OSTP), and private organizations, including the Howard Hughes Medical Institute, the Allen Institute for Brain Science, the Salk Institute for Biological Studies, and the Kavli Foundation. Coordination across NIH will be facilitated by the NIH Blueprint for Neuroscience, which brings together sixteen institutes and centers, and across Federal agencies

---

[5] According to the developer's paper (http://www.nature.com/nature/journal/v497/n7449/full/nature12107.html), "the term was an acronym to describe the Clear Lipid-exchanged Acrylamide-hybridized Rigid Imaging/Immunostaining/*In situ* hybridization-compatible Tissue-hYdrogel." However, due to the length of the actual phrase and because the method actually goes beyond the original acronym, we did not provide the phrase in the text above.

21

through the Interagency Working Group recently formed under the auspices of OSTP. In addition, the President also directed his Commission for the Study of Bioethical Issues to engage with the scientific community and other stakeholders to identify proactively a set of core ethical standards both to guide neuroscience research and to address potential ethical dilemmas raised by the application of neuroscience research findings. The President formally charged the Commission with this task on July 1, 2013, and the Commission will begin to deliberate the request at its public meeting on August 20, 2013.

The BRAIN Initiative can take many lessons from the success of the Human Genome Project, including the importance of computational sciences and sharing data widely, with appropriate privacy protections. High-resolution activity recordings and large-scale anatomical circuit reconstruction produce vast quantities of data, and making sense of multidimensional datasets generated through the BRAIN Initiative will be an enormous challenge. By one estimate, the human brain can produce in 30 seconds as much data as the Hubble Space Telescope has produced in its lifetime.[6] A joint NIH-NSF initiative on "Big Data Science" is underway, which will be important in addressing these data challenges. Furthermore, we do not understand the "neural code" through which the rate, timing, and patterns of activity in populations of brain cells represent and transform information. NIH and NSF have collaborated for almost a decade in computational neuroscience, which is essential to take on this challenge. Human brain imaging, including information from the ongoing NIH Human Connectome Project, will also complement the goals of the BRAIN Initiative.

To develop a blueprint for NIH's role in the BRAIN Initiative, NIH has established a group of 18 highly qualified external advisors, including 3 *ex officio* members from the Federal partner agencies. They have wide ranging expertise, including contributing to the development of

[6] Nature 499:274, 2013.

several of the transformative new techniques described today. This working group of the Advisory Committee to the NIH Director is expected to present interim recommendations to the Director late this summer and final recommendations are anticipated in the summer of 2014. The group is already interacting extensively with the NIH research community to develop a more complete scientific plan, anticipated next summer, that will: 1) identify high priority investments (*e.g.,* improving current tools, identifying new directions); 2) develop principles for achieving the goals (*e.g.,* the balance between small groups and large consortia); 3) suggest collaborations with foundations, industry, other agencies, and international programs; and 4) deliver timelines and milestones.

In basic neuroscience, as in other areas of research, NIH emphasizes investigator-initiated research, which engages the wisdom and ingenuity of the research community. Tool development empowers the research community by providing better, cheaper technologies and open-access databases. We are confident that the BRAIN Initiative will have a catalytic effect on neuroscience research in the coming years. In addition to the BRAIN Initiative, NIH will also continue to support research across the full range of basic and applied neuroscience.

There are excellent reasons for our confidence in the importance of the BRAIN Initiative as a critical aspect of research addressing brain disorders. Foremost, many diseases of the brain, including autism, dystonia, epilepsy, and schizophrenia, are fundamentally disorders of brain circuitry; and others, such as Parkinson's disease and Alzheimer's disease, cause symptoms by disrupting the performance of circuits as brain cells degenerate. Even with our limited understanding of brain circuits and the relatively imprecise technologies for interventions, therapies that compensate for malfunctioning brain circuits already produce remarkable results for some people. Deep brain stimulation, which uses chronically implanted electrodes, has proven to have long term benefit in clinical trials for Parkinson's disease and dystonia, and has

shown promise for Tourette's syndrome, depression, epilepsy, obsessive compulsive disorder,

chronic pain, and several other disorders. Cochlear implants send coded auditory signals that

brain circuits can interpret to restore useful hearing to thousands of people, visual prostheses

have restored rudimentary sight in pilot studies, and brain computer interfaces that monitor

signals from the movement control circuits of the brain have enabled people with paralysis to

move a robotic arm, using only their thoughts. Decades of pioneering NIH interdisciplinary

research and, more recently, extensive cooperation with DARPA, have been critical to these

advances in neural prostheses, some of which you will hear more about today. These advances

provide a glimpse of the future. Better understanding of brain circuits would not only improve

the sophistication of stimulation-based interventions and prosthetics, but also inform

development of treatments for the full range of brain diseases that affect circuits. Looking more

broadly, unanticipated benefits will likely flow from the BRAIN Initiative beyond neuroscience,

just as the Human Genome Project had entirely unexpected benefits. History has shown that the

most important outcomes of the BRAIN Initiative may well be those we have yet to imagine.

Department of Health and Human Services
National Institutes of Health
National Institute of Neurological Disorders and Stroke
Story C. Landis, Ph.D.

Dr. Story C. Landis began her appointment as the Director of the National Institute of

Neurological Disorders and Stroke (NINDS) on September 1, 2003. A native of New England,

Dr. Landis was awarded her B.A. degree in Biology, with highest honors, from Wellesley

College (1967), and her M.A. (1970) and Ph.D. (1973) degrees from Harvard University. After

postdoctoral work at Harvard University studying transmitter plasticity in sympathetic neurons,

she served on the faculty of the Harvard Medical School's Department of Neurobiology.

In 1985, Dr. Landis joined the faculty of the Case Western Reserve University School

(CWRU) of Medicine in Cleveland, Ohio, where she held many academic positions, including

Professor and Director of the Center on Neurosciences, and Professor and Chairman of the

Department of Neurosciences, a department that she was instrumental in establishing. Under her

leadership, the CWRU Department of Neurosciences achieved worldwide acclaim and a

reputation for excellence. In 1995, Dr. Landis was appointed as the NINDS Scientific Director,

and was responsible for the direction and re-engineering of the Institute's intramural research

program. Beginning in 1999, in conjunction with the leadership of the National Institute of

Mental Health (NIMH), she spearheaded a movement to bring a sense of unity and common

purpose to the numerous laboratories, in multiple NIH Institutes, that conduct leading edge

clinical and basic neuroscience research, which increased research cooperation and collaboration,

and resulted in the construction of the new NIH Neuroscience Research Center, Phase II of

which will be completed this fall. Since early 2007, Dr. Landis has also been Chair of the NIH

Stem Cell Task Force.

Throughout her research career, Dr. Landis has made many fundamental contributions to understanding the developmental interactions required for synapse formation, and has garnered many honors and awards. In 2002, she was elected President of the Society for Neuroscience, and served as President-elect until her appointment as the NINDS Director in September 2003. Dr. Landis is an elected Fellow of the American Academy of Arts and Sciences and the American Association for the Advancement of Science and the Institute of Medicine of the National Academy of Sciences.

Chairman BUCSHON. Thank you very much.

Now, I recognize Mr. McLoughlin for his testimony.

TESTIMONY OF MR. MICHAEL MCLOUGHLIN, DEPUTY BUSINESS AREA EXECUTIVE, RESEARCH AND EXPLORATORY DEVELOPMENT AT APPLIED PHYSICS LABORATORY, JOHNS HOPKINS UNIVERSITY

Mr. MCLOUGHLIN. Chairman Bucshon, Ranking Member Lipinski, Members of the Subcommittee, thank you for the opportunity to come and talk to you today and to tell you about some of the progress that we have made in the area of brain-controlled prosthetics.

This program was initiated in 2005 by DARPA to provide enhanced capabilities for soldiers who had experienced upper extremity amputations. We have also since included patient populations that are affected by spinal cord injury or other neurodegenerative conditions which prevent them from using their natural limbs.

The objective of this program was to develop—is to develop a prosthetic limb that really has all the capability of our natural limb system. And so the challenge is to provide a level of functionality that begins to rival that of what was lost due to the amputation.

In conducting this work, we have had to work with—had the fortune to work with multiple government agencies, including the NIH, who you just heard from, as well as a team of researchers across this country that have totaled over 30 different organizations that range from research groups doing basic research to very applied engineering and to work across those groups in order to solve this challenge.

So in other words, basically four major challenges that we are addressing here, the first one was to develop a prosthetic limb, as you see here, and that Sergeant Deslauriers is wearing that can mimic the function of the natural arm. And we had to do that in a form factor that matches the natural limb, so tremendous set of engineering challenges here.

The second challenge was to be able to control the limb. So we all do very complex things with our arms and we do it very naturally. We don't even think about it. For a prosthetic user, these become very difficult, requiring tremendous concentration. And yet our brains do it every day without thinking. So the major focus of our programs has been looking at direct interfaces with the brain in order to control the limb system.

The third area then is to provide sensation from the limb. So we can all utilize our limbs without looking at them. So I can reach out and grasp an object. I know where my arm is. I know what it is touching. A prosthetic user cannot do that. So what we are investigating is ways that we can feed information back to the brain to provide sensory perception.

We have already demonstrated that for amputees, that stimulation of the residual sensory nerves can provide very vivid sensation to the level of the patient will actually say I feel my finger, okay. I am not—I don't feel where you are touching it; I feel my finger that was lost. We are beginning now to explore how do we provide that same level of capability to somebody that has a spinal cord injury that we can directly input that information into the brain.

The last area is to provide a fundamental research capability that can live beyond just what we are doing in this program. It will provide a set of tools that can be used by researchers and developers of new medical devices, rehabilitative devices, in order to push the field of neuroscience forward.

I would like to now show a quick video.

[Video.]

This is Tim Heans at the University of Pittsburgh, one of our research participants. He was the first person to drive this limb using just a brain computer interface. Tim was injured in a motorcycle accident and is paralyzed from the neck down, and he is controlling his arm strictly by thinking about where he wants it to go. And so this is after about actually just about a day of working with the arm. And here you see him reaching out to one of the members of the research team, and when his girlfriend saw this, she said I want to try this. And so she got up and for the first time since his injury, Tim was able to actually reach out and physically interact with another human being. And this was a tremendous impact to Tim and to his girlfriend. And Tim, when you hear him talk, will actually say I will reach my arm out to touch her. So it gives you a sense of the meaning to these patients.

[The prepared statement of Mr. McLoughlin follows:]

Statement to the Subcommittee on Research and Technology

Mr. Mike McLoughlin, Deputy Business Area Executive

The Johns Hopkins University Applied Physics Laboratory

July 31, 2013

Chairman Bucshon, Congressman Lipinski, and Members of the Subcommittee. Thank you for inviting me to provide my views on the revolution in prosthetics. I will provide you my opinions and a short demonstration of what technology can provide for those who have lost a limb. The opinions stated are my own and do not necessarily reflect those of the Johns Hopkins University Applied Physics Laboratory or the Department of Defense.

The Defense Advanced Research Projects Agency initiated the Revolutionizing Prosthetics program in 2005 to provide expanded prosthetic options for Warriors who experienced traumatic upper-extremity amputations during contingency operations in Iraq and Afghanistan. After APL was awarded its contract to support the Revolutionizing Prosthetics program in 2006, we learned that many upper extremity amputees chose to either wear a body-powered split-hook or no prosthesis at all. The split-hook technology, which has not changed much since it was patented in 1912, and the other prosthetic arm components available in 2006, offered very little natural arm and hand function. Our task at APL was to develop a neurally-integrated yet modular anthropomorphic arm system with near-natural control. Today, I am happy to report we have met that challenge and have prototype systems that are providing great hope to people with arm amputations and to those that have lost the ability to control their limbs due to spinal cord injury, stroke, or neuro-degenerative diseases.

In addition to support from DARPA, we have been fortunate to collaborate with several government agencies, including the National Institutes of Health, Walter Reed National Military Medical Center, Uniformed Services University of the Health Sciences, Veterans Health Administration, US Army Medical Research and Materiel

Command and the Food and Drug Administration. This collaboration has been integral to the success of the dedicated team led by the Johns Hopkins University Applied Physics Laboratory and comprised of researchers from the Johns Hopkins University, the California Institute of Technology, the University of Southern California, Rancho Los Amigos National Rehabilitation Center, the University of Pittsburgh, the University of Pittsburgh Medical Center, the University of Chicago, Northwestern University, Hunter Defense Technologies, Blackrock Microsystems, and the University of Utah.

Our team strove to accomplish four major goals. First, we developed a prosthetic arm that was able to mimic the natural arm. The Modular Prosthetic Limb (MPL) is capable of articulating 26 joints, enabling it to do almost anything the natural arm can do. Unlike most current prostheses, the MPL is a completely integrated system that can be separated into three modules, full-shoulder, humeral or radial configuration; each including the wrist joint and hand. The MPL can be easily adapted for use with conventional body attachments. Many users have stated that the natural motion of the MPL is an extremely important aspect of the MPL resulting in a feeling of embodiment.

Our second objective was to provide a means for the user to naturally control the arm. To attain this level of control, the APL team initially explored peripheral nerve control options and showed that interfacing with muscles and nerves in the residual arm of research participants with an amputation allows them to achieve remarkable functional improvements with prostheses. Since 2010, we have focused on direct brain interfaces that have demonstrated amazing control capabilities for our research participants with tetraplegia. Magnetic resonance imaging (MRI) studies show that even years after an amputation or onset of paralysis, the areas of the cortex associated with motor control are still activated by the thought of moving a lost limb. The team's challenge was to access and interpret these brain signals in a way that would enable the research participant to move the MPL as they would move their natural arm.

Third, we are seeking to restore natural sensation via sensors on the prosthetic limb. The ability to perceive physical interactions with objects and sense the position of our natural limb without seeing it allows us to walk into a dark room and not fall or to reach into a handbag and grasp an object without seeing it. Just like the brain remembers how to move a lost limb, research suggests the areas of the brain involved in perceiving sensation are still active and capable of perceiving sensation. While much less is known about generating sensory perception, we do know that stimulation of sensory nerves years after an amputation can produce vivid sensations that are associated with the missing limb. Rehabilitative surgeries, such as Targeted Muscle Reinnervation (TMR), have provided the ability to stimulate peripheral nerves in residual limbs based on inputs from the MPL hand sensors; amputees have stated they can feel sensations in their phantom fingers as a result of these stimulations. In addition, research participants have stated they experience great relief from phantom limb pain after their TMR surgeries. With over 100 embedded sensors in the MPL, we have only begun to explore the possibilities of restoring lost perception and the sense of touch.

The final goal was to create a foundation of achievements that could be leveraged into future efforts to advance the capabilities of neuroprosthetics for the benefit of people with unique rehabilitation and restoration needs. This required a systems approach to integrating advances across multiple disciplines including: neuroscience, neurosurgery, biomedical engineering, electrical engineering, and mechanical engineering. It was through the systematic integration of all these advances that we were able to translate neuroprosthetic technologies into practical applications that met the Revolutionizing Prosthetics program goal of functional restoration.

During the course of the Revolutionizing Prosthetics program, we have made significant advances and learned important lessons. We developed advanced algorithms that can decode a paralyzed person's intent to move their arm and hand from their neural brain signals. As a result, research participants have the ability to move the MPL in a very intuitive way, often stating that they are just thinking about moving their own arm or simply thinking about manipulation of a target object. This control extends to the ability to form multiple grasps, an achievement that realizes the goal of dexterous prosthetic arm and hand control. Our TMR efforts have provided neurosurgery insights that include the benefit of preserving the nerves in the residual limb to provide the possibility of using signals from those nerves to control prostheses and provide sensory feedback to the user. These new surgical approaches could eliminate the need for secondary surgeries to address phantom pain. Another important advance was the development of miniature motors. Customized motors in the joints of the MPL allow for near-natural movements and deliver the high torque required for human-like strength in a small lightweight package.

In addition to advancing these engineering and clinical capabilities, APL created a virtual reality version of the MPL for use when a physical limb system is not available. The Virtual Integration Environment (VIE) is completely interchangeable with the MPL, providing the research community with a low cost means of testing brain computer interfaces. The VIE is being used to test novel neural interface methods, study phantom limb pain, and serves as a portable training system. We have already provided the VIE to multiple research groups at universities across the United States at no cost. By extending this open platform approach we can stimulate other research teams and foster increased interdisciplinary collaboration.

It is my experience that the technologies developed under the DARPA Revolutionizing Prosthetics program have had a profound positive impact on the research participants. Whether due to amputation or a spinal cord injury, users of the MPL system have had very positive reactions to being able to move a limb for the first time since their injury or illness. We have seen individuals with limb loss - often years after their injury - learn to move the prosthetic arm in minutes. Johnny Matheny, who lost his arm to cancer, underwent TMR surgery and tells of being able to feel his hand again and about relief from phantom limb pain. Air Force Master

Sergeant Joe Deslauriers, lost both legs and his left arm in Afghanistan, and has demonstrated how advanced control algorithms enhanced his ability to use a prosthetic arm. Research participants with paralysis have been able to physically interact with others via the MPL driven by their brain signals. For the first time in the seven years since his accident, Tim Hemmes, an individual with tetraplegia, was able to hold his girlfriend's hand using the MPL. Similarly, Jan Scheurmann was able to feed herself without assistance for the first time in over ten years using the MPL. These, and other research participants, have become integral members of the team; and their bravery and dedication is an inspiration to the rest of the research team.

While the team has accomplished much since 2006, many opportunities lie ahead of us to provide greater independence and quality of life to people with disabilities and the elderly. By using advanced algorithms to control robotic devices, I believe we can significantly reduce cognitive burden for those using assistive devices to accomplish everyday tasks. I believe it is essential to continue efforts to develop neuroprosthetic devices that will allow for natural control of replacement limbs. While today we are focused on research participants with amputations and paralysis, the insights we are gaining promise to help the elderly and those that suffer from amyotrophic lateral sclerosis (ALS), stroke, multiple sclerosis, and traumatic brain injury. Finally, I feel it is important for these efforts to lead to the development of a clinical standard of care that is viable for transition to the home and workplace, enabling individuals with disability to lead more productive and independent lives.

Again, Mr. Chairman, thank you for this opportunity to inform Congress on the practical benefits that are coming out of the long-term Federally-funded research on advanced prosthetics and related fields, and I look forward to answering your questions.

Michael P. McLoughlin
Deputy Business Area Executive, Research and Exploratory Development

EXPERIENCE

Mike McLoughlin is the Deputy Business Area Executive (DBAE) for the Johns Hopkins University Applied Physics Laboratory (JHU/APL) *Research and Exploratory Development Business Area.* He leads world-class teams with a broad range of disciplines to develop, integrate, and transition innovative systems using advanced biomedical technologies to meet the needs of the military. In 2009, Mike assumed leadership responsibilities for Defense Advanced Research Projects Agency's (DARPA) Revolutionizing Prosthetics Program (RP) at JHU/APL and is leading efforts to transition use of neutrally integrated RP technologies to human subjects. His technical experience includes development of rapidly emerging technologies and sensor systems, technology transfer, program management, signal processing, applied engineering and technology development, aerosol science research, and microbiological laboratory analysis.

Mike began his career at APL in 1985, where he has held progressively responsible line and program management positions (Deputy BAE, Branch Manager, Program Manager, and Systems Engineer). During his tenure at APL, he has had significant roles in the Biomedicine and Homeland Protection Business areas; established and managed new programs, and built the technical organizations required to support these fast growing enterprises. In addition, Mike teaches both Program Management and Systems Engineering at the Johns Hopkins University Whiting School of Engineering.

EDUCATION

☐ Bachelor of Electrical Engineering, University of Delaware, 1983
☐ Master of Electrical Engineering, University of Delaware, 1984

Mr. McLOUGHLIN. I am very fortunate today to have with me Master Sergeant Deslauriers, who has been one of our research participants, and I would like to give him a moment to tell you about his experiences working with the arm.

TESTIMONY OF U.S. AIR FORCE MASTER SERGEANT JOSEPH DESLAURIERS JR.

Sergeant DESLAURIERS. Again, Chairman, I would like to echo my thanks for the opportunity to speak with the panel today.

It has been about a year since I have been working with the limb after my injury on September 23, 2011. When you lose three limbs at once, it is very difficult to figure out how you are going to interact with the world around you now. I was—I am a husband, I am a father. How am I going to hold my child? How am I going to interact? And when the opportunity came up to work with the gentleman from Johns Hopkins University, I kind of jumped at the chance to aid in the research of the arm, and it was an honor for me to help with the advancement of prosthetics for upper limbs.

Working with the arm, it has been amazing because the limbs that we have now for upper extremities are not very versatile. They don't have many degrees of movement. I will get a wrist turn and maybe a pinch, but with this, I can open my hand. I can rotate my wrist. I can grab something. And it is amazing to have something that you can manipulate with your residual limb and eventually with your brain. It gives you that confidence and that independence to get back into the work field and continue to serve your country in whatever manner be so. Thank you.

Chairman BUCSHON. Thank you very much. And thank you again for your service to your country. It is very much appreciated.

I now recognize Professor Raichle for five minutes to present his testimony.

TESTIMONY OF DR. MARCUS RAICHLE, PROFESSOR OF RADIOLOGY, NEUROLOGY, NEUROBIOLOGY AND BIOMEDICAL ENGINEERING, WASHINGTON UNIVERSITY

Dr. RAICHLE. Chairman Bucshon, Ranking Member Lipinski, and Members of the Committee, thank you so much for inviting me to participate in this hearing to discuss future prospects for neuroscience research.

Having been involved in neuroscience research for the past 45 years, and must say that I am—my life has been—I have been very fortunate to experience an absolute revolution in the way we think about and look at the human brain. And this of course came about in the 1970s when x-ray computed tomography, CT, the CAT scan was introduced. It not only changed the world of neurology in which I work, but also it promoted thinking along the lines of other ways in which to obtain images of organs of the body and particularly the human brain.

The first to appear on the scene was positron emission tomography or the PET scan which was invented in our laboratory in the early 1970s and followed thereafter by the development of magnetic resonance imaging. And both of those techniques have matured tremendously over the intervening years and are providing us with spectacular information on the human brain and health and disease across the lifespan from premature infants to the end of life, valuable insights that were unanticipated when I got into this business.

This of course is—involve the efforts of a wide range of highly skilled technical people in areas of physics and engineering and chemistry and computer science. But to me one of the great advances in all of this was creating the interface of this technology to the study of the human brain. And therein it called upon and benefited enormously from an understanding of how to describe human behavior. This is no mean task and it involved people at the outset beginning to study issues of language and linguistics and cognitive psychology and it was instrumental in the development of the field of cognitive neuroscience, which I think is a marvelous demonstration of integration of talent across multiple levels that is necessary if you are going to make any progress in this endeavor.

Much of the imaging that one sees in the now—something on the order of 17,000 papers in the world literature on fMRI and another 14,000 involving PET, what one sees is often traditionally a way of looking at the brain, of asking you to do something and comparing it to you are not doing it and seeing what lights up. And so you can see this in scientific journals in Newsweek and Time magazine and probably on TV on occasion.

And this dominated the story for quite some period of time and is still an important part of this, but there came a realization along the way that these changes that we observe, that which is occurring in my brain as I talk to you and in your brain as you listen to me, are small changes in the background of enormous activity. Your brain on average is about two percent of your body weight and yet it consumes 20 percent of the body's energy budget. So if you are just being a neural economist, you would say we better find out about what this is all about.

And how this has evolved has been quite remarkable in the sense that this ongoing activity is noisy, and for a long time we just threw it away. Scientists like to get rid of noise in their data. And then there came the realization that this noise is deeply interesting, and from it, we can determine remarkable insights in terms of how the brain is organized in carrying on its activity regardless of whether you are sitting here in this room sleeping, driving your car, or whatever.

So this has been a paradigm shift in the way we operate and think about this, this whole idea of intrinsic activity, and its importance is, I think, immense in terms of understanding the diseases of the nervous system because if you are going to do that, you are going to have to understand what the nervous system is actually doing and what it is devoting its efforts to.

Now, what—much of what I have said and which I think about of course is of great interest to neuroscientists writ large, but, as was posed to me in the questions for this committee, what about the man in the street, the person that is concerned about a disability, a history of Alzheimer's and their family? And it is incredibly prevalent in mine. And what I can say is that from this work what has emerged is the ability to predict the onset of disease because what can't be replaced must be prevented.

So in the case of Alzheimer's, the ability to anticipate the onset of the disease by many years using imaging materials which, if I had had more time, I would love to show you, but I think the issue of using these biomarkers of disease to anticipate the onset of symptoms by years allows us to think creatively about preventing the disease before it take its toll. Thank you very much.

[The prepared statement of Dr. Raichle follows:]

TESTIMONY

of

Marcus E. Raichle, M.D.
Washington University School of Medicine
St Louis, Missouri

to the

Committee on Science, Space, and Technology
Subcommittee on Research and Technology
U.S. House of Representatives
31 July 2013

Good morning Mr. Chairman and members of the subcommittee. Thank you for the opportunity to provide testimony on the frontiers and challenges of human brain research, including its potential and limitations in curing brain diseases. My name is Marcus Raichle. I am a neurologist and the director of the Neuroimaging Laboratories in the Mallinckrodt Institute of Radiology at Washington University in St Louis where I am Professor of Radiology, Neurology, Neurobiology, Psychology and Biomedical Engineering. I am a member of the National Academy of Sciences, the Institute of Medicine, the American Academy of Arts and Sciences and a Fellow of the American Association for the Advancement of Science.

Human brain research has advanced tremendously over the past 40 years beginning with the introduction of X-ray computed tomography or CT in 1973 by Godfrey Hounsfield, an English electrical engineer working at EMI, Ltd. CT obtains its information by passing X-ray beams through the brain at many different angles to measure its density. CT not only revolutionized the way we look at the brain by providing the first true 3D brain images in living subjects but also stimulated the development of two other imaging techniques that together have provided unprecedented images of the anatomy and function of the human brain in health and disease. The first of these was positron emission tomography or PET which allows us to measure brain function in terms of its chemistry, circulation (i.e., blood flow) and metabolism using unique, cyclotron-produced radioisotopes. The other technique, introduced shortly after PET, is magnetic resonance imaging or MRI which provides superb anatomical images of the brain as well as measurement of its ongoing function. MRI obtains its information by measuring the properties of atoms in a strong magnetic field. The technology behind each of these techniques is most remarkable and continues to evolve particularly in the case of PET and MRI.

While technological developments involving physicists, engineers, chemists and computer scientists have been critical to the development of PET and MRI for human brain research the success achieved in using these techniques to image the human brain in health and disease has required extremely important input from clinical and basic neuroscientists as well as behavioral scientists schooled in techniques required to quantitatively measure human behaviors. A prime example of the collaborative nature of this work was the first major study of language organization in the normal human brain obtained with PET in 1988. This was the culmination of over 15 years of work by a multidisciplinary team of investigators whose talents ranged from computer science and image processing to linguistics and cognitive psychology.

Growth in functional imaging research of the human brain has been exceptional. Since its introduction in 1993, functional MRI or fMRI as it is best known has accounted for over 17,500 in the world's scientific literature along with an additional 14,000 papers using PET. The thirst for information about the brain, particularly the human brain, is universal and imaging for better or worse has been used by many as a medium for the discussion.

Functional brain imaging has followed a long tradition in neuroscience: studying neuronal responses to stimuli and during task performance. The resulting images, which now appear in thousands of scientific papers as well as in the popular press, routinely show areas of the brain which 'light up' as tasks are performed. These images represent the *difference* between performing a task and/or viewing a stimulus, and a control condition, which could be as simple as lying quietly in a scanner with your eyes closed. While the changes that are observed in these *difference images* have been immensely valuable in showing us the complex network of brain areas involved in particular tasks they obscure the fact that most of the activity of the human brain is ongoing at all times regardless of what one is doing. This has come to be known as the brain's intrinsic activity. How do we know this?

This discovery of the importance of the brain's intrinsic activity arose from a consideration of the cost of brain function. As adults we invest 20% of our body's energy budget in brain function, an organ that represents only 2% of our body weight. This means that the cost of brain function is 10 times that expected on the basis of its weight. When comparisons were made between this enormous ongoing cost of brain function and the additional cost of task performance it was realized that the latter was a trivial addition, usually just a few percent locally and not detectable when looking at the overall cost. From this work it became apparent that if we were to understand normal brain function in health and disease it will be critical to increase our understanding what the brain was really doing as represented by its intrinsic or ongoing activity.

fMRI has been invaluable in opening the door to an understanding of the brain's intrinsic activity. When one examines the brain with fMRI what is striking, but for a long time ignored, was the fact that the images contained a great deal of 'noise' (i.e., seemingly random signal changes). What scientists often do in the presence of noise in their data is to average across many data points thereby eliminating the noise. For many years this is exactly what was done with fMRI data until it was realized that this 'noise' contained a remarkable amount of information about the ongoing organization of the brain. In fact, this discovery has been so remarkable as to cause a paradigm shift in the way in which functional brain imaging is used to study the human brain. So-called 'resting state' studies where subjects simply lie quietly in an MRI scanner with their eyes closed have become a standard component of imaging research. It is not only important in revealing the large scale organization of the human brain but also a valuable tool in studying the effect of disease where task performance can sometimes be difficult if not impossible for patients.

Human brain imaging as it is now being performed with MRI and PET is providing an increasingly detailed understanding of the human brain: how its component parts are organized into a functioning system, how they develop and change with age, how the brain is wired together, how the metabolism of the brain varies with the needs of its component parts and how all of these aspects of the organization and function of the brain are affected by disease. The pursuit of this broad agenda of research is exemplified by the recently initiated, NIH sponsored *Human Connectome Project*.

One of the great challenges for neuroscience is now making sense of this enormous influx of new data. Lurking beneath the surface of the fascinating new images of the human brain is the need to make sense of the brain signals that generate these images. How do changes in brain circulation and metabolism, the basis of our imaging signals, relate to changes in brain electrical activity? Traditionally, brain electrical activity, particularly the spiking activity of neurons, has been considered the primary signal of brain activity. It is now clear that the spiking activity alone is insufficient to understand brain function. Other electrical activity occurring in the membranes of brain cells, both neurons and supporting cells, contribute to brain function and brain imaging signals. Furthermore, the complex metabolic machinery within brain cells of all types is setting the stage for functional brain activity that is being programmed at very basic levels by the genes expressed within brain regions and with specific cells types and their processes. We must, therefore, invigorate a dialogue among scientists across levels of analysis from cell biology, genetics and neurophysiology to human brain imaging. This will challenge the comfort level of many but is necessary if we are to make progress. Schooling young investigators to think broadly and in new ways will be essential to progress. New tools will, of course, aid in this endeavor just as CT, PET and MRI created a revolution in the way we view the human brain. But tools alone, in my estimation are not sufficient if not accompanied by an integrated sense of how we should approach our understanding of brain function in health and disease.

Much of what I have said relates to how neuroscientists are thinking about brain function but what about the average American who has or worries about someone afflicted by a brain-related disease? In providing an answer to this question it is appropriate to say that anticipating brain disease is critical. No better example exists than in the case of Alzheimer's disease where biomarkers derived from imaging will play a critical role in understanding the disease years in advance of symptoms and evaluating therapies before symptoms appear and irrevocable damage has occurred. When damage has occurred, such as in stroke, understanding why some recover and other do not, an active area of imaging research, will lay a much better foundation for rational approaches to rehabilitation. Finally, in diseases such as depression and Parkinson's disease imaging has provided critical information about the circuits involved and has allowed neurosurgeons to place stimulating electrodes in the brain that greatly reduce otherwise intractable symptoms much as cardiac pacemakers aid in correcting abnormal heart function.

Finally, it must be said that fascination with how the brain works captures the imagination of scientists and lay persons alike and in so doing enriches discussions of how we behave has human beings.

Thank you for your attention.

Marcus E. Raichle, a neurologist, is a Professor of Radiology, Neurology, Neurobiology and Biomedical Engineering at Washington University in St Louis. He is a member of the National Academy of Sciences, The Institute of Medicine and the American Academy of Arts and Sciences and a Fellow of the American Association for the Advancement of Science.

He and his colleagues have made outstanding contributions to the study of human brain function through the development and use of positron emission tomography (PET) and functional magnetic resonance imaging (fMRI). Their landmark study (Nature, 1988) described the first integrated strategy for the design, execution and interpretation of functional brain images. It represented 17 years of work developing the components of this strategy (e.g., rapid, repeat measurements of blood flow with PET; stereotaxic localization; imaging averaging; and, a cognitive subtraction strategy).

Another seminal study led to the discovery that blood flow and glucose utilization change more than oxygen consumption in the active brain (Science, 1988) causing tissue oxygen to vary with brain activity. This discovery provided the physiological basis for subsequent development fMRI and caused researchers to reconsider the dogma that brain uses oxidative phosphorylation exclusively to fuel its functional activities.

Finally seeking to explain task-induced activity decreases in functional brain images they employed an innovative strategy to define a physiological baseline (PNAS, 2001; Nature Reviews Neuroscience, 2001). This has led to the concept of a default mode of brain function and invigorated studies of intrinsic functional activity, an issue largely dormant for more than a century. An important facet of this work was the discovery of a unique fronto-parietal network in the brain that has come to be known as the *default network*. This network is now the focus of work on brain function in health and disease worldwide. Of particular interest is the fact that this newly-discovered network is a primary target of Alzheimer's disease.

In summary, the Raichle group has consistently led in defining the frontiers of cognitive neuroscience through the development and use of functional brain imaging techniques.

Chairman BUCSHON. Thank you very much.

I now recognize Professor—Dr. Robinson for five minutes to present his testimony.

TESTIMONY OF DR. GENE ROBINSON, DIRECTOR, INSTITUTE FOR GENOMIC BIOLOGY, SWANLUND CHAIR, CENTER FOR ADVANCED STUDY PROFESSOR IN ENTOMOLOGY AND NEUROSCIENCE, UNIVERSITY OF ILLINOIS, URBANA-CHAMPAIGN

Dr. ROBINSON. Good morning, Chairman Bucshon, Ranking Member Lipinski, and Members of the Subcommittee. I would also like to thank you for the opportunity to provide testimony today on the frontiers in human brain research and the importance of an interdisciplinary and interagency approach to neuroscience.

Today, I will use an example from my laboratory's research on honeybees to address the importance of basic research on brain and behavior. It is necessary to understand how healthy brains work in order to find treatments for the many devastating brain disorders that afflict our society. This involves basic research on animal models, the type of science that is championed by the National Science Foundation. From this work, we can generate hypotheses for what changes occur in a dysfunctional system and then test possible interventions for these disorders.

If I may have the first image, please?

[Slide.]

Honeybees are famous for their highly structured division of labor. Some bees take care of the baby bees while others forage outside for nectar and pollen. In addition to this highly structured organization, there is also a great deal of flexibility. Bees can switch between jobs according to the needs of their colony. This raises the question how can a brain that is the size of a grass seed produce such complex behavior? What does this say about our brains?

To address this question, we developed a couple of new research tools. One is a new system of tracking bees with radiofrequency ID tags developed in my laboratory by retired businessman and current citizen scientist Paul Tenczar to help us study behavioral activity.

The second tool is a device to study brain activity that comes from genomics, which is a new science that studies the assemblage of all of our genes. We suspected that switching from one job to another might involve reprogramming the bees' brains for the new job. This led us to interdisciplinary research from behavior to genomics with funding from NIH and USDA to sequence the bee genome. We were surprised to find that the way this reprogramming occurs is that the genome actually is very sensitive to the environment and in a very dynamic way.

When a bee responds to events in the hive, thousands of genes in the brain change their activity and then the behavior changes. It is as if the genes are blinking on and off like Christmas lights, changing the amount of the brain's proteins that they make. It turns out that in addition to bees, other species, including birds, fish, mice, and humans also have dynamic genomes in their brain.

Last year, I co-chaired a special meeting of the National Academy of Sciences and the Canadian Institute for Advanced Research to explore the human health implications of this discovery of the dynamic genome. The conference imagined a new interdisciplinary collaboration among psychologists, sociologists, political scientists, neuroscientists, and geneticists to understand how the experiences of childhood adversity affect the brain and predispose for certain types of brain disorders. The lesson here is that an insight from basic animal research is helping to address the critical question in human health.

It will take the integration of a variety of types of research on both animals and humans to reach a complete answer, including research funded by the NSF Directorate for Biological Sciences and the NSF Directorate for Social, Behavioral, and Economic Sciences. The BRAIN Initiative similarly needs to commit to an effective blend of basic and applied research to provide more opportunity for transformative discoveries.

The bee story also illustrates that some animals are ideally suited for the pursuit of very specific questions, sometimes even better than the traditional workhorses of the laboratory, the fruit fly or the mouse. Neuroscientists actually have known this for a long time. The humble squid essentially launched the modern era of neuroscience because its nerve cells are so big that their activity could be studied even with the primitive techniques of the 1940s. The research undertaken as part of the BRAIN Initiative should likewise benefit from a broad research agenda of model animals and model behaviors.

Understanding how the brain works represents a formidable challenge to our collective ingenuity and dedication. With this challenge comes great opportunity to increase our understanding of brain and behavior to improve our health and the functioning of our society. We must remember that basic science research is called basic not because it is simple but because it provides the foundation for innovation.

Through the united and creative efforts of biologists, mathematicians, engineers, physicians, and other explorers of the brain, big brains or little brains, we must and we will find the answers that we need. Thank you.

[The prepared statement of Dr. Robinson follows:]

TESTIMONY
of
Gene E. Robinson, Ph.D.
Swanlund Chair of Entomology and Neuroscience
Director of the Institute for Genomic Biology
University of Illinois at Urbana-Champaign

to the

Committee on Science, Space, and Technology
Subcommittee on Research and Technology
U.S. House of Representatives
July 31, 2013

Good morning, Chairman Bucshon, Ranking Member Lipinski, and members of the Subcommittee. Thank you for the opportunity to provide testimony today on the frontiers in human brain research and the importance of an interdisciplinary and interagency approach to neuroscience. My name is Gene E. Robinson and I am the Swanlund Chair of Entomology and Neuroscience and the Director of the Institute for Genomic Biology at the University of Illinois at Urbana-Champaign. I am a member of the US National Academy of Sciences and currently serve on the Advisory Council of the National Institute of Mental Health.

Today I will address three topics: first, the importance of basic research on brain and behavior; second, the wisdom of studying a diverse set of animal models; and third, the power of interdisciplinary research, which is essential for building new tools to study the human brain.

A few years ago, the National Science Foundation (NSF) sponsored a workshop that I chaired to address the challenges of 21st century biology. Our report, published in 2010 in *BioScience*, concluded that, "Addressing the challenges of 21st century biology requires integrating approaches and results across different subdisciplines of biology…as well as technologies, information, and approaches from other disciplines…" This applies to many areas of biology, including neuroscience, and in particular, the recently announced Brain Initiative, that is, the Brain Research through Advancing Innovative Neurotechnologies Initiative. The BRAIN Initiative needs to develop a broad and inclusive agenda that funds basic research on brain and behavior, both in humans and in a variety of animal species.

Why is a broad approach necessary? What are the benefits of studying a wide array of species in our efforts to understand the human brain? One of the goals of the BRAIN Initiative is to understand how the brain produces human behavior, with all of its complexity and potential for disorder. We are fortunate that the diversity of animal life on the planet provides us with many potential models for aspects of human behavior, so long as we have the knowledge to recognize and take advantage of them. This approach of exploring and capitalizing on the resources provided by nature falls perfectly within the mission of NSF. NSF supports a wide scope of basic science on brain and behavior that provides the breadth of knowledge necessary for continued advancement of the field of neuroscience.

It is necessary to understand how healthy brains work in order to cure the many devastating brain disorders that afflict our society. This involves basic research on animal models—the type of science championed by the NSF. The role of NSF in meeting societal challenges is to support basic research—research that examines how a healthy system functions and adds to our knowledge of how living things work. This knowledge makes it possible to generate hypotheses that describe what changes occur in a dysfunctional system, and to propose and test possible interventions for those disorders. The process is interconnected, interdisciplinary, and progressive. Let me give you one example from my own research to illustrate the benefits of integrating research on the brain with research on behavior in an interdisciplinary manner, as well as the synergy between basic science and its sometimes unexpected applications.

I study social behavior, specifically how it arises in nature and what mechanisms govern it. I use honey bees to address these questions. The reason I use honey bees is that they live in one of the most complex societies on the planet, with tens of thousands of individuals involved in intricate forms of communication and division of labor. Intriguingly, they produce all this complex social behavior with a brain the size of a grass seed! How can such a tiny brain produce such complex behavior and what does this say about our own brains?

One question my laboratory has addressed is how do bees know what job to perform in their hive. There are about a dozen different jobs that bees perform, including feeding the baby bees, foraging for nectar and pollen from flowers, and turning nectar into honey. Bees divide labor in a very organized fashion, with different groups specializing in the different jobs. But bees don't do these jobs like little robots; rather they adjust their behavior to the needs of the whole group. When a hive of bees loses some of its foragers, others will drop what they're doing and start to forage. Thanks to a new system of tracking bees with Radio Frequency ID tags developed in my laboratory by retired computer entrepreneur and current citizen scientist Paul Tenczar, we can now automatically detect these adaptive behavioral shifts, enabling us to more easily explore the underlying neurobiological questions.

One intriguing aspect of job-switching in bees is that they do it without receiving commands from centralized control. Neither the queen nor any other individual directs the actions of the rank and file worker bees, but everyone in the beehive does what needs to be done. Each bee is able to synthesize information about the environment inside and outside the hive, along with internal cues about its physiological state, to appropriately direct its own behavior. We suspected that this might involve reprogramming the bee's brain to perform the different job, but we needed new tools to monitor brain activity.

In the early 2000s, with the advent of more advanced genome sequencing technology, we pushed to sequence the honey bee genome, the assemblage of all the genes, in order to develop powerful new tools for brain analysis. Fortunately the NIH's National Human Genome Research Institute at the time was considering additional species for sequencing in order to better understand the human genome and the honey bee was selected because of its compelling social behavior. The United States Department of Agriculture also contributed to this effort because of the vital role bees play in pollinating our nation's food and fiber crops, contributing approximately $20B per year to our economy.

45

With new tools in hand we obtained a grant from the NSF Frontiers of Science program, which was designed to promote interdisciplinary research. My laboratory performed a series of experiments that explored the relationship between job shifts in the beehive and changes in brain gene activity, which leads to changes in how much of the brain's proteins are produced.

We found that the brain of the bee is indeed reprogrammed to perform a different job, but the way this reprogramming occurs was a big surprise. Not only does the genome provide a script for building and operating the brain; when it comes to behavior, the genome also improvises—it is sensitive to the environment and alters the activity of genes in a dynamic way. When a bee detects a decrease in the number of foragers in its hive, thousands of genes in its brain change their activity, and this causes the bee to start to forage. The bee's experience is embedded in its genome in the brain so that it can change its behavior appropriately.

It turns out that bees are not the only ones with dynamic genomes in their brains. Birds, fish, mice, and other animals also have been found to exhibit dynamic brain genomes. In addition, as expected when a feature of biology is similar in many different organisms, humans also appear to exhibit the same dynamic brain genome. It is much more difficult to study this phenomenon in humans than in animals, and it likely never would have been done without animal discoveries paving the way.

Last year I co-chaired a special meeting of the National Academy of Sciences and the Canadian Institute for Advanced Research to explore the human health implications of this discovery. Social scientists have carefully documented the developmental and health consequences of early exposures to adversity, and now they badly want to know how the experiences of childhood adversity get "under the skin," into the body's systems that influence vulnerability and resilience. The conference imagined a new "science of adversity," with a new partnership among psychologists, sociologists, political scientists, neuroscientists and geneticists.

The question of biological embedding—how social influences are perceived, processed, and ultimately transformed into signals inside brain cells--is one of the most important questions in neuroscience, with profound health and public policy implications. It is clear that early adversity changes behavior, learning capacities, and social functioning, but how this happens –how the brain develops differently under adversity—can only be studied in animals in a basic research framework. Our research on honey bees helped initiate this line of research, but it will take the integration of this and many other research efforts to reach a complete answer, including research on animals funded by the NSF Directorate for Biological Sciences and research on humans funded by the NSF Directorate for Social, Behavioral and Economic Sciences.

My laboratory's research on honey bees shows the value of combining the power of new technology with knowledge derived from basic research on both brain and behavior. The BRAIN Initiative similarly needs to commit to an effective blend of basic and applied research, to increase the opportunity for transformative discoveries. The initiative will likewise benefit from the selection of experimental models and behaviors that provide illuminating contexts in which to apply them. However, neuroscientists are increasingly relying on the study of just a few species to understand behavior and brain function. These classic model organisms, including the fruit fly and mouse, do offer experimental advantages. They are easy to breed in the laboratory and they are easy to use for many types of genetic experiments to learn the functions of different genes. As attractive as these advantages are, the use of only a few model organisms is unnecessarily limiting. Many aspects of biology are the same across species, but each species has unique characteristics as well; to distinguish between these two possibilities, multiple species must be studied and compared. The unique features of some organisms offer research opportunities that more traditional study organisms do not, often because they represent an extreme of a biological property of interest. Studies that make strategic use of well-chosen, and diverse, animals models can have tremendous impact on a field. Neuroscientists have long known this—the lowly squid essentially launched the modern era of neuroscience because its nerve cells are so big that their activity could be studied even with the primitive techniques of the late 1940s. Because an increasing number of species have had their genomes sequenced over the past few years, there are more choices than ever before for high-powered molecular analyses of the brain.

Our 21st century biology report also concluded that, "biologists need devices to continuously record the activity of cellular components as they interact naturally in living cells." This recommendation has been embraced by the BRAIN Initiative, and future technological innovation will be central to uncovering the workings of the human brain. Here, again, NSF's contributions will be vital because of its tradition of encouraging and facilitating interdisciplinary approaches to integrate engineering, computer science, physics and chemistry. The new tools to record brain activity are most easily developed by individuals who combine knowledge of physics or math, or expertise in the applied skills of computer science or engineering, with understanding and appreciation of the challenges and technological needs of biology. The University of Illinois at Urbana-Champaign provides an excellent example of an environment that fosters this type of interdisciplinary work. Modern research institutes such as the Institute for Genomic Biology and the Beckman Institute for Advanced Science and Technology bring together top engineers and biologists in a spirit of open communication and collaboration. This atmosphere has led to remarkable innovations. For example, an NSF-sponsored partnership between the University of Illinois and the Cray Corporation built Blue Waters, one of the world's most powerful supercomputers, the computational capacity of which will vastly improve our ability to model the most complex biological systems, including the human brain.

Understanding how the brain works represents a formidable challenge to our collective ingenuity and dedication. It is important that we consider carefully how to best direct our efforts and resources to meet this challenge, united by our common interest in improving the health and structure of our society. I appreciate the opportunity to be here today with you and the committee to discuss this important topic. We must remember that basic science research is called "basic" not because it is simple, but because it provides the foundation for innovation. I am confident that this initiative will bring great improvements to our understanding of the human body, the brain, and our health by promoting the continuation of impressive work in our university research centers and government laboratories, in partnerships with private organizations, and enabled by funding from government agencies. Through the united effort of biologists and mathematicians, engineers, physicians, and other explorers of the brain, both big brains and little brains, we must –and we will-- find the answers we need.

Gene E. Robinson obtained his Ph.D. from Cornell University in 1986 and joined the faculty of the University of Illinois at Urbana-Champaign in 1989. He holds a University Swanlund Chair and is also the director of the Institute for Genomic Biology and director of the Bee Research Facility. He served as director of the Neuroscience Program from 2001-2011, leader of the Neural and Behavioral Plasticity Theme at the Institute for Genomic Biology from 2004-2011, and interim director from 2011-2012. He is the author or co-author of over 250 publications, including 26 published in *Science* or *Nature*; pioneered the application of genomics to the study of social behavior, led the effort to gain approval from the National Institutes of Health for sequencing the honey bee genome, and heads the Honey Bee Genome Sequencing Consortium. Robinson serves on the National Institute of Mental Health Advisory Council and has past and current appointments on scientific advisory boards for companies with significant interests in genomics. Dr. Robinson's honors include: University Scholar and member of the Center of Advanced Study at the University of Illinois; Burroughs Wellcome Innovation Award in Functional Genomics; Founders Memorial Award from the Entomological Society of America; Fulbright Senior Research Fellowship; Guggenheim Fellowship; NIH Pioneer Award; Fellow, Animal Behavior Society; Fellow, Entomological Society of America, Fellow, American Academy of Arts & Sciences; and member of the US National Academy of Sciences.

Chairman BUCSHON. Thank you very much.

And thank you all for your testimony. It is fascinating. I am really going to be interested in seeing where the questions lead us today. It is going to be a fascinating discussion.

I want to remind mebers of the committee that the rules limit questioning to five minutes. And at this point I will recognize myself for five minutes.

There was a study, Dr. Raichle, in National Geographic about caffeine. I don't know if you saw that one about people waking up in the morning just as a sideline and studying the brain flow—colored brain flow of people that are decaffeinated and people that have caffeine, and it is true you do need your cup of coffee in the morning if you are chronically a caffeine user. It showed that.

Dr. RAICHLE. Fortunately, I had mine.

Chairman BUCSHON. There you go. It was a fascinating, fascinating study.

Along the similar line, you mentioned that if we could image diseases earlier in lives, we may predict what might be the future. I mean we have diseases like Huntington's chorea, for example, and we do know genetically what will happen. Has that disease or any other like that been helpful? And anyone else that wants to comment can also. Dr. Raichle? I mean is there—is that what you are talking about?

Dr. RAICHLE. Not Huntington's in particular. The one that stands out in my mind, of course, is Alzheimer's because of the enormous effort to look at the changes early on realizing that they do occur 15, 20 years before the onset of symptoms.

In a slide that I was hoping to show you but didn't the project known as the Dominantly Inherited Alzheimer's Network, which is studying these rare genetic variations that guarantee you will get Alzheimer's disease, they are rare but they enormously informative, you can predict in an individual when they are going to get the symptoms. So studying them 15, 20, 25 years beforehand, you can begin to categorize the changes in—of the pathology like amyloid plaques and the changes in metabolism, the brain atrophy that precede the onset of symptoms by many years.

This opens up an opportunity to understand how the disease evolves but it also opens up the opportunity of slowing it down or preventing it. And in the case of Alzheimer's, simply slowing it down has an enormous benefit to family and to the individual and to the economic cost of that terrible disease.

Chairman BUCSHON. Do you have anything to add to that, Dr. Landis?

Dr. LANDIS. In Huntington's disease, there are longitudinal studies that have been tracking people who are known to be gene-positive, and looking both at imaging parameters and psychosocial parameters, and we now have the same kind of understanding that is involved in Alzheimer's. Before the motor symptoms appear, it is very clear that there is quite a long prodromal period. And just like for Alzheimer's, were there neuro-protective therapies that had been identified, you could in fact treat patients before there is enough destruction of neurons to actually see motor symptoms. Similar studies are underway for Parkinson's disease.

Chairman BUCSHON. Thank you.

Mr. McLoughlin, your team is composed of engineers, medical doctors, surgeons, and scientists working closely together. These are individuals that would not normally work together. What elements are required for successful interdisciplinary approach, I mean, in your view?

Mr. MCLOUGHLIN. Okay. I think there is basically four elements that are present here that are all very important. First of all, we are able to leverage decades-long basic research in the brain. And so we have research members on our team that have been supported by NIH and others that have spent years understanding how to take a set of neural patterns in the brain and understand what the intent was, how to form the hand. And that was a obviously very important piece of this.

The second component of this was advances and technology outside the field of neuroscience. So, for example, in the back of the Joe's hand here is a small processor which is essentially the same thing that most of you have in your smartphones right now. So it allows us to do all the very complex analysis in that very small package. So it can be self-contained, portable, lightweight.

The third component then was we—DARPA recognized that there was a need, so I am old enough to remember Neil Armstrong walking on the moon and that program was driven by a singular objective, which is put a man on the moon and return him safely. And this project has a similar objective that unifies the team—a very diverse team. So we have basic researchers through very applied engineers that are all very much focused on the fact of developing a prosthetic arm that works like our natural arm. And it is—and I can state that very concisely, very simply. In everything we do on the program is towards that objective and it doesn't matter where—you know, if you are working in a basic laboratory or you are doing CAT–CAM designs of mechanical devices somewhere.

The fourth very critical element is the environment which we develop this in. So we very early on made a very conscious decision that we would maintain an open architecture to the system so that while we have our research team working on this, we have other research teams that are currently using pieces of the technology that have come out of this program, imported it into their laboratories, and have done that very, very easily.

So we allow researchers to come in, modify the system, connect their own things to it so it makes a very easy, open platform so that researchers aren't having to constantly reinvent things in order to work in this area. So we put all those things together and provided, you know, the environment, you know, the basic science, and that singular drive in order to pull this whole set of players together, which we have had over 30 different organizations involved in this program.

Chairman BUCSHON. Great. That sounds like it has been a fairly cohesive effort towards a singular goal, and that seems like maybe your most important message.

I am going to recognize now the Ranking Member, Mr. Lipinski, for his questioning.

Mr. LIPINSKI. Thank you, Mr. Chairman. I want to—before I begin, I want to ask unanimous consent to enter into the record the opening statement by Ranking Member Johnson.

50

Chairman BUCSHON. Without objection.
[The prepared statement of Ms. Johnson follows:]

PREPARED STATEMENT OF FULL COMMITTEE
RANKING MEMBER EDDIE BERNICE JOHNSON

Thank you Chairman Bucshon. I'm really delighted to be here this morning. In my hometown of Dallas, the Center for Brain Health at the University of Texas at Dallas is doing important research on brain disorders and injuries and contributing to the Administration's BRAIN Initiative. I have taken a number of people to the Brain Health facility so we could talk to the researchers and learn more about their work.

Before I entered public service, I was a psychiatric nurse at the VA Hospital in Dallas. This was at a time when many of our young men were returning from Vietnam seemingly whole on the outside, but suffering from acute and long-term mental health challenges that we only recently came to understand as post-traumatic stress disorder. Today, because of the life-saving measures that we have been able to implement in the field, thousands of young men and women have survived serious injuries in Afghanistan and Iraq and returned to their families. But many of them, and many more without any visible scars, suffer terribly from traumatic brain disorder and PTSD.

The research supported by federal agencies such as NSF, NIH, and DARPA is essential to increasing our understanding of the human brain. We need to better understand when things go wrong, such as in PTSD and drug addiction, so that we may develop more effective treatments. But it's hard to determine when things have gone wrong if we don't fully understand the normal functioning of a healthy brain. Because the National Science Foundation is not limited by examining specific pathologies or applications, it is particularly well suited to asking and answering fundamental questions about normal brain function. With this freedom, NSF can support research such as Dr. Robinson's work on understanding the social behavior of honey bees. As Dr. Robinson's work evolved from his basic questions about honey bee behavior, the applications to human neuroscience became evident and NIH also began to fund him. This is the way it should work. As we put neuroscience in context at today's hearing by focusing on applications, we should not forget the foundation of basic research on which these advances are built or the agency that is the leader in supporting such basic research.

Dr. Robinson, I'm sorry for putting you on the spot, but your work in particular illustrates another important point. Five years ago you published an NIH funded study on the Effects of Cocaine on Honey Bee Dance Behavior. If I were to look just at that title in order to judge the merits of your research, I might dismiss it as unworthy of taxpayer support. But I have confidence in NSF's and NIH's merit review process, a process that has become recognized worldwide as the "gold standard" for merit review. As a result, I have no doubt this is a serious study with real implications for understanding human addiction, an important issue in neuroscience. I also wonder about the significance of this work to better understanding honey bee colony collapse disorder that threatens agricultural production worldwide. I hope you will have the opportunity during Q&A to enlighten us on this fascinating research.

Thank you all for being here this morning and I look forward to your testimony.

Mr. LIPINSKI. Thank you. I want to thank all of our witnesses and especially thank Master Sergeant Deslauriers for his service to our country.

I want to start with Dr. Landis. What are the distinctive roles of the Federal partners in the BRAIN Initiative? And the second part is who is managing the program ensuring that the work is coordinated?

Dr. LANDIS. That is an excellent question. There are three Federal partners that are currently involved: NIH, NSF, and DARPA. There is an interagency working group on neuroscience that has been set up to look at interests of many more Federal agencies in brain research, and they have written a report which is not yet public which has recognized that the BRAIN Initiative or projects

51

like that are a critical part not just for those three agencies but for all agencies.

There are commitments that are made for Fiscal Year 2014 from the three agencies. NIH is in the process of planning what those initiatives will look like. We expect a report early in September. And on that committee—NIH committee sit ex officio members from DARPA and from NSF. NIH has been involved in the NSF planning. And so I anticipate that, based on the missions of those agencies, we will end up with a very complementary and integrated program.

DARPA, as you have heard, has mission. We want to fly to the moon. We want to create a prosthetic arm. NIH has interests in integrative science, mammalian—not just mammalian but many models. And NSF has—brings to the table engineering, mathematics, and other approaches.

So we believe that through collegial interaction and participation in the planning efforts that this will be a well-managed project. But it isn't yet launched so we will see as it goes forward.

Mr. LIPINSKI. All right. Thank you. I want to move to Dr. Robinson. In your opening statement you brought up how important it is to have an integrative approach to research topics like this, and you point out the considerable resources that the University of Illinois can bring to bear from neuroscience to social science to the computing power of the Blue Waters computer.

So I would like to ask you for your vision of what is possible over the next ten years of this initiative over these disparate fields. I know it is a huge question but just to give us some sense of what types of questions you think we will be able to answer ten years from now that we can't today.

Dr. ROBINSON. I can give you one general vision and that has to do with an approach in science is to really understand a particular phenomenon. One needs to be able to do two things. One needs to be able to observe it under natural conditions and then one needs to be able to manipulate it. So a lot of the BRAIN Initiative is geared toward developing new tools to be able to visualize the activity of a real live active brain and see it in action when it is responding to changes in its environment, when it is called upon to organize a particular activity.

And so there is a great deal of excitement about the development of sensors that are at the nanoscale. We have some superb engineers at the University of Illinois who are getting mobilized to work on these now thanks to the BRAIN Initiative, the sensors that work at the nanoscale that will be possible then to record the activity of an active brain, and then in turn to be able to use that same inroad into the brain to be able to stimulate particular parts of the brain, particular circuits to get more specific cause-and-effect relations.

And then finally, tying that altogether will be really high-powered computer models, the kind that Blue Waters will be able to do to be able to understand the phenomena, decompose it into single-unit-level understanding, as well as the whole level.

Mr. LIPINSKI. I understand that you did a very good job of putting out there for us what needs to come together in all this. Is there anything that you would expect? What kind of—you know, just look out there and say what would you like to solve? What do you think we can solve? What types of questions or problems or issues, is there anything that you have in mind?

Dr. ROBINSON. We spoke today. Several people mentioned how the brain is organized hierarchically. There is different levels of organization. You have whole brain and you have brain regions, you have circuits, and then there are the individual neurons. We badly need to understand the relationship of those units to each other, those levels of organization to each other. How do individual neurons orchestrate their activity to create a circuit? How do the circuits then form a brain region that is functional? And then of course the whole brain.

I take inspiration in framing this question from the beehive, no surprise, where we have similar questions. So you have a fully functioning colony and we need to understand how the behavior of individual bees gives rise to the whole colony and how the brain inside the brain—how the brain inside the bee gives rise to the colony and the gene inside the brain inside the bee inside the colony. So it is a Russian dolls nested-level sort of approach, and that is exactly what any complex system has. And the challenge is to decompose into the functional levels and then understand the relationship between those functional levels.

Mr. LIPINSKI. All right, thank you.

Chairman BUCSHON. Thank you very much.

I now recognize Mr. Hultgren for five minutes.

Mr. HULTGREN. Thank you very much. I really appreciate you all being here. And this is so interesting. Hang on one second. My phone is—Gina is taking it out. Thank you. I bumped something and I apologize. Bad timing.

It is—this is so interesting for me and I really appreciate you all being here and want to see this as a start. And I want to thank the Chairman and Ranking Member for their efforts in starting this discussion and really figuring out where we can take this from here. Brain science and brain injury and illnesses impact so many people. We may see just the human toll through Alzheimer's, Parkinson's, but also for young people. Some of the challenges we are seeing there as well, even at very young ages with some educational challenges with brain science and—or brain diseases that we don't fully understand.

So I just want to thank you so much for being here. Thank you for your work.

I do want to talk briefly on some issues that I am focusing on right now. And, Dr. Raichle, I know you mentioned our brains take up about two percent of our body weight but use about 20 percent of the energy. One of the things—and I am so thankful for Dr. Robinson and Blue Waters and what they are doing at the University of Illinois.

What we have seen China now surpassed us in computing power and I am encouraging—we have got legislation that we have introduced to push our own abilities into exascale computing and recognizing how important computers are going to be for us to be able to continue brain research. And so I wanted to just get your thoughts on that. It is interesting. The human brain can do more parallel computations per second than our fastest supercomputer while riding on the energy required for a dim light bulb, just amazing.

But there are really incredible challenges that we face as well. I know that we can reduce the amount of energy needed for these exascale computing challenges but also some of the parallelism challenges are going to be there.

So I wondered if—I know the Human Brain Project is one of European Commission's Future & Emerging Technology flagship projects. The goal for that is to reconstruct the brain piece-by-piece using supercomputer-based models and simulations. I know these models offer the prospect of a new understanding of the brain and its diseases leading to completely new computing and robotic technologies.

I wondered, Dr. Landis, and then also Professor Raichle if you could talk just briefly about the European Commission. They have announced this ten-year plan with funding levels of, I think, it is $1.19 billion. What are your thoughts on this project? Why have they taken this approach? And do you think if you could get some thoughts, do think this is the correct approach and is it something we can learn from here as well of planning towards the future?

Dr. LANDIS. So the two projects, the BRAIN Initiative and the European Human Brain Project are actually quite different in the approaches that they are taking and very complementary. I just spent the last two days at a planning meeting, an NIH planning meeting for the BRAIN Initiative. And what became very clear at that meeting was that in order to come up with reasonable models of how brains function, you really need to have data about the system itself and that models in the absence of the data about how the brain works really are not going to be terribly useful.

So you can think of our BRAIN Initiative as producing tools that would allow us to collect those data and that the Europeans will be going ahead trying to create models perhaps in the absence of all the data that they need.

Now, China has also—is also embarking on a brain project that seems to be the next big thing, and of course you have mentioned the concerns about Chinese investments in computers. We in the States, I think, in the neuroscience community are concerned about investments that other countries are making in neuroscience and other biomedical disciplines and about brain drain. And it is hard not to have young scientists see opportunity where funds, investments are going up instead of down.

Mr. HULTGREN. We do this. I am going to run out of time. And so I do want to follow up with all of you if that is all right. I have a lot of other questions and things, but I want to just spend my last minute or so with Master Sergeant.

First of all, thank you so much for your service. I was just struck as you are talking of your commitment to continue to serve in new ways, and I just think that is amazing. And I would just ask you, and Mr. McLoughlin as well, your thoughts. You talked about quality of life for our women and men who have been injured in service that, but I wondered also if you could talk briefly if this could potentially have application as well in areas of high danger dealing with explosives and things and what is happening with that and if you see much of a future there? Certainly, we want to help people who have been injured but the best thing would be to prevent the injury in the first place, and if that very dangerous job to be done by something—a machine like this. I wonder if you could talk briefly about that.

Sergeant DESLAURIERS. Yes, sir, absolutely. Well, I am coming up on 16 years in February so I have been doing this long time.

Mr. HULTGREN. Thank you.

Sergeant DESLAURIERS. And we kind of grew into it and, you know, the idea where it came about, you know, since 2000—I mean since 9/11. So, the quality of life for us since then, I kind of have a perspective of both sides being an amputee and then also being an explosive arms disposal craftsman where I see, you know, I can use this on a daily basis but then I could also use that on a robot to take that—take it out from a vehicle, send it down range, and I can take apart and IED just as easily as I would be doing it with my own hands.

Mr. HULTGREN. It is amazing.

Sergeant DESLAURIERS. I just tried that one out for the first time today and I was amazed. And it opened my eyes up to the program aside from the prosthetic side and seeing the other applications of the MPL. So it is not only going to be for the quality of life of amputees in the future not only just military but also civilian and then with the application of putting it into the field for future use and saving lives.

Mr. HULTGREN. Great. Well, again, my time is expired. Thank you, Chairman. But I just want to again thank you so much. Master Sergeant, thank you for your work on this and your continued commitment to see advancement in this and protect future soldiers as well. So thank you all so much and look forward to continuing the conversation and taking this forward. Thank you so much.

I yield back.

Chairman BUCSHON. Thank you.

I now recognize Mr. Peters for five minutes.

Mr. PETERS. Thank you very much, Mr. Chairman. And thank you, Master Sergeant, not just for your service but what you are going to help teach other people who have been similarly affected. And thank you for that, too.

Two lines of questions maybe for Dr. Landis. You mentioned how the BRAIN Initiative can take lessons from the successful human genome project, which we in San Diego feel a particular connection to. And you include the importance of widely sharing data. So I am curious about what policies, including data management and access, you think are in place or need to be in place to make sure that the data generated from the BRAIN Initiative can be shared across disciplines and ultimately into the private sector?

Dr. LANDIS. So the issue of data sharing has become increasingly important as scientists collect larger and larger data sets. They need to be available and accessible to appropriate scientists to analyze. We have excellent examples with the human genome project and also with ADNI, Alzheimer's Disease Neuroimaging Initiative, which posts on websites for people to see as soon as the data are collected. The human Connectome Project is posting data quarterly. We anticipate that that data sharing and mechanisms to permit it will be an integral part of the BRAIN Initiative.

And part of the meeting that I just attended was dealing with what kinds of data do we need to share and what kinds of repositories do we need and how we have appropriate access? So it is very much on the minds of the committee.

Mr. PETERS. Top of mind in the BRAIN Initiative. That is the place to be.

Dr. LANDIS. And you do have a representative on the planning committee from San Diego——

Mr. PETERS. Right. I appreciate it.

Dr. LANDIS. —not a Representative, a scientist from your district.

Mr. PETERS. And then my second question has to do with the outputs from this in addition to the research itself, in particular training opportunities. Anyone—this could be anyone—training opportunities, an initiative, whether NIH has a role in training undergraduates and graduate students in other fields? And then kind of implications for new curricula or degree programs that we might want to institute for the next generation of brain scientists? And maybe, Dr. Landis, you could start and anyone else could respond.

Dr. LANDIS. So for training, part of the NIH mission is not only to discover fundamental knowledge and apply that knowledge but also to train the next generation of biomedical investigators. And we feel very strongly at NIH that that training begins at the level of college. And if you want to have first-rate investigators who are well-trained, you need to engage their interests in college and then to be able to frame appropriate training programs in graduate school and postgraduate. So we are very much committed to that.

In terms of the BRAIN research initiative, the discussion has been that if one of the most important things that we can do in the BRAIN Initiative is to analyze data and put together an understanding of how thousands or millions of neurons are interacting to create behavior, we really need to engage scientists in cross-disciplinary training that would take mathematicians, statisticians, and others, computational people to work hand-in-hand with investigators who are doing the wet bench work. So we talked about possible—expanding present training programs.

And I will cede to someone else.

Mr. PETERS. Okay. Anyone else want to comment on that? No? Well, I would say again, thank you, Mr. Chairman, for the hearing and thanks to the witnesses for being here. Again, in San Diego this is one of the cornerstones of our economy is the relationship between basic science research and in particular healthcare and brain research. So we are excited about it and hope to be participants and beneficiaries and wish you the best.

Chairman BUCSHON. Thank you.

I now recognize Mr. Collins for his questions.

Mr. COLLINS. Thank you, Chairman.

Dr. Landis, Buffalo, New York, is a hotbed for multiple sclerosis. As we know, MS is a genetically based, European-based auto-immune disease, and whether it is western New York or Australia, New Zealand, Europe, that is where we find it. So we are a hotbed for that and there has been a lot of drug development for relapsing-remitting, no question about it, but when it comes to secondary progressive MS, which you mentioned, which is where I would like to go, that is debilitating and an awful situation.

You mentioned that the NIH has been working on something which would be, you said, slowing the progression. I am just curious. I know of one drug out there that works with a very tiny subset of secondary progressive patients. I know of another, a microparticle immune—you know, stimulant that is looking to stop the progression. And I am just curious. Could you give me some more information on what you were referring to as something that was slowing the progression?

Dr. LANDIS. So I should have specified that I was referring to relapsing-remitting. We do not have treatments for progressive multiple sclerosis. And I would be pleased to get back to you with an answer for the record that would summarize the research in this area that NIH is conducting and what are the most promising avenues. We recognize that this has been an underexplored area. It is complicated. Not a lot of patients, but for the patients who have it, it is truly devastating. So I will get back to you with an answer.

Mr. COLLINS. Well, I think it is fairly well understood that almost every relapsing-remitting patient——

Dr. LANDIS. Becomes eventually—

Mr. COLLINS. —someday they will unfortunately move into secondary progressive at which point that is not a good day for them or their families. I do think the Fast Forward Fund, which I am sure you are familiar with, has worked on several. I do know there is one drug, MIS416, which is a microparticle immune stimulant that is in Phase IIB trials that has promise——

Dr. LANDIS. Right.

Mr. COLLINS. —on secondary progressive MS, but everywhere in western New York, especially, you know, as people look out 20 years and that is the typical relapsing-remitting time frame that it is not—so I am glad to hear you are working on it and I would very much like to know because I——

Dr. LANDIS. And if you would like to come and visit the intramural program, we have several investigators working on MS and would be pleased to have you come and meet with them and see the labs and some of the kind of approaches we are taking.

Mr. COLLINS. I definitely would like to take you up on that. It is an important part of what is going on in western New York and thank you very much.

Dr. LANDIS. Yes.

Mr. COLLINS. Mr. Chairman, I yield back.

Chairman BUCSHON. Thank you.

I now recognize Mr. Schweikert for five minutes.

Mr. SCHWEIKERT. Thank you, Mr. Chairman. Have you ever shown up at something and it turns out to be just fascinating?

And, Master Sergeant, thanks for spending time with us. I know sometimes sitting down, you know, in this sort of formal body can be a little nerve-racking and it is truly appreciated.

And let's start, Dr. Landis, and this may be one for everyone. First off, on diseases of the brain, let's focus on Alzheimer's, whether it be plaque or neurons that die and there are firing issues, where are we in the genetic modeling? And some of this is going to tie back to some things Dr. Robinson was saying. Where do you believe we are on understanding the map?

Dr. LANDIS. So we have identified a number of genes which are dominantly inherited and cause Alzheimer's. Dr. Raichle discussed one of them; there are several others. We have other genes which have been shown to increase risk. The most prominent of these is ApoE4. If you have two alleles ApoE4, you have a significantly greater risk of getting Alzheimer's. But there are still significant investments that can be made in this area, and one of the major projects from last year's special Alzheimer's money was to take $25 million of the $50 million and invest it in a better understanding of risk factors for Alzheimer's.

Mr. SCHWEIKERT. Okay. In that line, Dr. Robinson, was I listening to you properly, that some of your research or the externality of your research is the ability of observing the turning on and off of certain genetic mapping? Am I listening properly?

Dr. ROBINSON. Yes, that is correct. So there are tools now to be able to look at the activity of genes. Now, these tools are best deployed in animal models and they need increased sophistication to be able to be used in humans, but the initial insights can be gained from the animal models.

Mr. SCHWEIKERT. And are you—do you tie sort of your research into the mapping data now? Or are you still moving mostly, you know, moving from bees now to the next level of animal models?

Dr. ROBINSON. So we are collaborating in a broad network to be able to generalize the results from animals to the study of adversity, the program that I mentioned where we are looking at how—basically how the social environment, how do experiences "get under the skin" to affect biology, predispose for certain diseases.

Mr. SCHWEIKERT. Okay. I am going to do one bump and then back—Dr. McLoughlin, where are we technologically right now on nerve actually communicating with an interface? And where is it going right now and how much world and outside and private, you know, research are you seeing on innovation? I mean what is moving right there?

Mr. MCLOUGHLIN. Okay. So the state-of-the-art right now is that we have—so we currently have two patients that have been implanted with arrays. In these are arrays that have 100 electrodes so, you know, we talk about trillions of neurons, so we are seeing very, very small populations of neurons. And so we can—with current technology we can put up a couple hundred electrodes in the brain right now, fairly close to the surface. And with those signals, we are able to do very high-level control of the arms, so reach out, grasp objects, do the, you know, types of things that we normally do.

Mr. SCHWEIKERT. And where I was going—and forgive me, I don't remember the reference, but earlier this year, I thought there was some excitement because of some nano sensors that were being tested? And you may have to help me out on this one. And that actually was the direction that that technology was supposed to go.

Mr. MCLOUGHLIN. Yes, so I think that—so that is where we are today. And the challenges that we have are—today is that those electrodes tend to degrade over time, so after a couple of years, the response goes down. So the exciting thing in some of these nanotechnology arrays, use of growth factors so that the nerves will actually—rather than pulling away from the electrodes, it will actually grow into the electrodes so that we will—I see within, you know, the next five years or so that we see next-generation array systems coming out that instead of working for a couple of years will have the potential to work 10, 20, or 30 years in the human brain.

Mr. SCHWEIKERT. Okay. And I am going to—well——

Dr. LANDIS. If I could just add electrode manufacture is one of the initiatives that has come up repeatedly in the planning sessions for the BRAIN Initiative that we need better ways to record from more neurons over a longer period of time with more fidelity. And I—we don't know what is going to be recommended but——

Mr. SCHWEIKERT. And are you finding research both in this country and around the world, both private and public in that area?

Dr. LANDIS. I am—there is interest in this but it is pretty clear that this is a very tough area. You are talking material science, you are talking about connections, you are talking about radio communication of these rather than wires. And I think significant Federal investment in this area would make a huge difference in encouraging both investigators and the academic and private sector to engage.

Mr. SCHWEIKERT. I am over my time. Thank you, Mr. Chairman.

Chairman BUCSHON. Thank you very much.

Before I conclude today's hearing, I would like to thank and recognize Melia Jones. Where is she? She is back there. Raise your hand. I thank her for her work on this Subcommittee for the past two years and wish her all the best with her future endeavors. The committee hates to lose her but our loss, I guess, is Texas A&M's gain. And again, thank you very much for your service to the committee.

I would like to thank the witnesses for their valuable and very fascinating testimony and the Members for their questions. The record will remain open for two weeks so some Members may submit some questions for a written response and additional comments. And I think we could go on for a long time on this subject. It is very fascinating.

So the witnesses are excused and the hearing is adjourned.

[Whereupon, at 12:16 p.m., the Subcommittee was adjourned.]

Appendix I

Responses by Dr. Story Landis

Questions for the Record
House Committee on Science, Subcommittee on Research and Technology
Hearing on "Frontiers of Human Brain Research"
July 31, 2013
Dr. Story Landis

1. **Pediatric low grade astrocytomas (PLGA) are slow growing children's brain cancer. Current treatments for these low grade gliomas, which include chemotherapy and radiotherapy, is invasive, highly toxic, and often life threatening. Progress for these treatments is hampered by the need to increase access to tissue samples, unlock some basic research challenges such as the lack of a mouse model, and to accelerate the basic and clinical research on the disease. How could the Brain Research through Advancing Innovative Neurotechnologies (BRAIN) Initiative help to advance research on PLGA slow-growing children's brain tumors?**

Although it is difficult to foresee the direct impact of the BRAIN Initiative on Pediatric Low Grade Astrocytoma (PLGA), knowledge gained about how neurons and neural circuits function and the interactions of neurons and glial supporting cells could yield valuable insights. In addition, data sharing and analysis tools and policies developed through the BRAIN Initiative could influence research on PLGA in useful ways.

Ongoing research supported by the National Institute of Neurological Disorders and Stroke (NINDS) and the National Cancer Institute (NCI) is currently increasing our understanding of the basic biology of PLGAs and paving the way for the development of new treatments. NINDS-funded researchers are studying the genetic and cellular changes in tumor cells and in the tumor microenvironment that contribute to the formation and maintenance of PLGAs. Alterations to the BRAF gene (either mutations or fusion of the BRAF gene with other genes) are common in PLGAs. NINDS and NCI are supporting the development of preclinical models that will allow identification of promising candidate treatments for PLGAs with BRAF gene alterations, which have been targeted in therapies developed for other cancers. The NCI-supported Pediatric Preclinical Testing Program developed a model for a BRAF mutation-positive subtype of PLGA and identified a targeted therapy, selumetinib (AZD6244), for additional research. NCI also supports several projects that focus on obtaining high-quality biospecimens and sharing of tissues for research purposes. As a partner within the International Cancer Genome Consortium, NCI continues to collaborate with German pediatric brain tumor research colleagues who are collecting and analyzing PLGA tumor tissue samples and using them to generate PLGA genomic sequence data. In addition, NCI supports the Children's Oncology Group (COG) Biopathology Center, the largest pediatric specimen bank in the country. An example specific to PLGAs and other brain tumors is a COG protocol focusing on collecting and storing blood and brain tumor tissue samples from children with brain tumors treated at COG institutions. The initiative provides long-term storage of these specimens and makes them available to qualified researchers to study the biology of pediatric brain tumors.

2. During the hearing we heard how understanding the brain is an interdisciplinary endeavor. Education clearly plays an essential role in preparing scientists to study the brain. What can be done, both at the K-12 level and postsecondary level, to ensure researchers and other scientists are prepared to work and think in an interdisciplinary framework?

Science education is not just important for future scientists. All of us face personal and social issues for which we would be better informed with an understanding of basic scientific principles. NIH provides many free resources for science teachers to engage students, and a wealth of educational material for public access on its website. NIH has contributed expertise in other ways as well, including close cooperation with the Smithsonian in developing an extensive new exhibit on the genome that opened recently. In 2011, the NIH Blueprint for Neuroscience issued a Request for Applications for Blueprint Neuroscience Research Science Education Awards after gathering suggestions from the public and from experts on what further role the NIH might play in developing neuroscience-related materials for K-12 education. The program required applicants to present innovative, creative plans for improving science knowledge and enthusiasm for science among students and teachers. NIH funded meritorious, peer reviewed proposals from eight teams of scientists and educators, each of which included plans for program evaluation. Those programs are now ongoing.

At the undergraduate level, many of our best colleges and universities do an excellent job of engaging the natural curiosity of students about the brain, but not all students have comparable opportunities. For this reason, the NIH Blueprint supports the Enhancing Neuroscience Diversity through Undergraduate Research Education Experiences (ENDURE) program. ENDURE aims to raise interest and opportunities in neuroscience research for individuals who are typically underrepresented in the field. ENDURE forges partnerships between research-intensive institutions and institutions with a substantial enrollment of neuroscience majors from diverse groups. At post-graduate levels, many NIH individual and institutional training programs emphasize the importance of interdisciplinary training and provide opportunities for research experiences that will encourage this outlook.

Across the Federal Government, NIH is also working with other agencies under the Interagency Working Group on Neuroscience (IWGN), established by action of the National Science and Technology Council (NSTC), Committee on Science, to coordinate activities in neuroscience research with a focus on identifying significant transformative opportunities of national importance. One specific goal of the IWGN is to improve our understanding of learning and cognition and applying that to improvements in education and other related areas. The work of the IWGN is anticipated to be publicly-released later this year.

3. I am concerned that we have reached a situation where there is more data being generated from experiments than our capacity to synthesize and understand. Could you elaborate on the importance of algorithms and analytical tools to help us understand the science of the brain? How much progress have we made in this area, and where should we focus limited funding resources?

Early in the discussion about the BRAIN Initiative, many neuroscientists emphasized the importance of having appropriate analytical tools in place and specific testable hypotheses about how brain circuits work as the rate of data accumulation increases. Neuroscience has a long history of using computational and analytical methods to understand experimental data. These methods have progressed as neuroscience itself has advanced, from calculations on hand cranked calculators more than 50 years ago that confirmed hypotheses about how nerve cells generate electrical impulses, through modern computer modeling that assesses whether specific hypotheses of how the brain develops and brain circuits work fit the experimental data. Thus, NIH recognizes the importance of not just generating data, but also analysis and synthesis to understand data, and support for these aspects of research is integral to NIH's funding for neuroscience.

The NIH BRAIN working group, a working group of the Advisory Committee to the NIH Director charged with developing the scientific plan for the BRAIN Initiative at NIH, is paying close attention to data analysis issues, and NIH has several general policies and programs that will be relevant to this issue as the initiative moves forward. On July 29, the NIH BRAIN working group held a meeting in Boston with the scientific community on "computation, theory, and big data," to ensure that these issues are incorporated in its final plan for the BRAIN Initiative. Also relevant, the NIH Blueprint for Neuroscience supports the Neuroscience Information Framework, which enables discovery and access of public neuroscience research data and tools worldwide through an open source, networked environment. For several years, NIH and NSF have jointly supported an initiative on computational approaches to neuroscience. The Human Connectome Project also illustrates the NIH emphasis on sharing data and analysis capabilities. More generally, NIH's National Center for Biotechnology Information (NCBI) provides databases that are essential tools for scientists throughout the world, and NCBI's experience as a major force in the development of modern genetic analysis will serve the BRAIN Initiative well. The NIH Common Fund also supports a program in Bioinformatics and Computational Biology, and recently launched a new initiative on Big Data to Knowledge, as well as collaborating with NSF in this area.

As mentioned above, NIH is also working with other agencies under the IWGN, established by action of the NSTC, Committee on Science, to coordinate activities in neuroscience research with a focus on identifying significant transformative opportunities of national importance. Another area of focus is encouraging public access to, and sharing and preservation of, neuroscience and related behavioral data, along with the development of relevant data infrastructure and analysis, visualization, and modeling tools, techniques, and methodologies. The work of the IWGN is anticipated to be publicly-released later this year.

4. In your opinion, are there any fields of brain science research that are neglected or underdeveloped? In particular, are you paying enough attention to rare brain disorders that only affect a few individuals, but have a greater potential to inform us about brain science research?

NIH places a high priority on research for rare disorders, both because of the enormous collective impact of these diseases on people and their families and because of what we can learn from rare disorders about normal biology and more common diseases. The value of studying rare disorders is nowhere more evident than in neuroscience, where classic studies of even single patients have yielded important insights that stimulate entire research areas. In the past, neurologists often inferred the functions of particular areas of the brain from patients with lesions in those brain structures. In the modern era, scientists frequently study genetic as well as physical lesions, that is, mutations that cause rare inherited diseases. These gene studies reveal information about the normal brain and common diseases, as well as about the rare disorders themselves. Studies of rare inherited subtypes of Alzheimer's and Parkinson's disease, for example, have revolutionized research on the common versions of these disorders, and Rett Syndrome is yielding valuable information about autism. Because of the importance of rare disease research, NIH funds these studies throughout its investigator initiated research programs and also through special programs targeting rare disorders, such as the Rare Diseases Clinical Research Network, led by the Office of Rare Diseases Research in close cooperation with the appropriate NIH Institutes.

With regard to neglected or underdeveloped areas of neuroscience, the NIH system of investigator-initiated research engages the entire scientific community in seeking out neglected problems and unmet opportunities for progress. The progress of science generally and the availability of a plethora of new techniques to study the brain enables the scientific community to generate many outstanding proposals to address key questions across widely diverse areas of neuroscience. NIH funds as many of these meritorious proposals as resources allow. To supplement the investigator-initiated avenue for research, when NIH recognizes that particular diseases are receiving less attention than warranted by their public health impact or the unique scientific opportunities that they present or special opportunities arise that are not well served by the investigator initiated grant mechanisms, we also solicit research proposals to address those gaps.

5. Do private foundations have an easier time funding high-risk high reward research in the area of brain science? Is it fair to say that the NIH funding mechanism is biased toward picking grants that are 'safer' in its aims? Please explain.

NIH has a long record of successfully supporting high risk, high reward research, as is evident from the 138 NIH-supported researchers who have received Nobel Prizes. In recent years, NIH has been concerned that intensified competition for funding might disadvantage innovative research and has taken several steps to address this. Following an external examination of its peer review practices, NIH revised the applications and review process. Shorter proposal formats now reduce emphasis on detailed description of proposed methods, and reviewers score proposals separately on specific criteria of innovation and significance. NIH also offers smaller, shorter duration grants, including R21 exploratory/developmental grants that specifically encourage proposals with potentially ground breaking impact and less preliminary data. NIH also created the Pioneer Awards, New Innovator Awards, and Transformative R01s, which are specifically designed to support high risk, high reward research and innovative investigators. Neuroscientists have done especially well in competition for funding in these NIH-wide programs. Neuroscience institutes have also supported the EUREKA (Exceptional, Unconventional Research Enabling Knowledge Acceleration) Awards. Some of the pioneering research that undergirds the BRAIN Initiative was supported through these innovation-directed programs.

NIH works closely with private organizations that support research, including innovative studies, but private groups cannot support the breadth of basic research and training that NIH does. Private organizations have used a variety of strategies to encourage high risk, high reward research, from goal-directed prizes to supporting investigators based on track record rather than for specific projects, but it is not apparent whether any of these strategies are consistently more effective than the NIH model. In fact, many prizes and investigator-focused awards from private organizations go to researchers who have received support from the NIH to develop their ideas. The BRAIN Initiative provides a good example of how NIH can collaborate with the private sector. The Kavli, Allen, and Howard Hughes foundations, and the Salk Institute, among others, support complementary research. With respect to disease-focused research, private disease organizations, especially those for rare disorders, often engage the interest of researchers and fund these investigators to obtain preliminary data that enables them to compete successfully for traditional NIH R01 grants. NIH frequently collaborates with private groups to organize workshops with the scientific community on particular diseases.

6. **How will you be able to quantify how well the BRAIN Initiative has achieved its goals and aims? What metrics or other means will you use to judge the effectiveness of the initiative?**

NIH has charged the Advisory Committee to the Director's BRAIN working group with developing a multi-year scientific plan for the BRAIN Initiative, which is to include timetables, milestones, and cost estimates. As part of the process, members will consult the scientific community, patient advocates, and the general public. The working group is expected to produce an interim report containing recommendations for high-priority FY 2014 investments in September and a final report in the summer of 2014. Based on these planning recommendations, NIH will develop metrics to judge progress. However, in crafting goals, it will be important to keep in mind the long term goal, which is not easily captured by quantitative metrics—understanding how a brain circuit performs its computations is more important than, for example, how many nerve cells a researcher can monitor simultaneously. Historians of science tell us and economists have confirmed that basic research yields substantial returns on investment, but the most important benefits are often not predicted or easily quantifiable.

7. **The creative drive and engine for American Science is the individual investigator. This is one of the reasons we have the peer review process at agencies like NSF and NIH. Why then do we need the BRAIN Initiative? If we have faith in the peer review system, won't these projects get funded anyway?**

The purpose of the BRAIN Initiative is not to detract from creative drive of the independent investigator, but rather, to enable more of these investigators to be capable of generating a deeper understanding of how the brain works. For this reason, the early focus of the BRAIN Initiative will be on the development of new tools and technologies that enable more researchers – especially individual investigators – to have access to critical resources necessary for advancing their own research aims. The Human Genome Project is an excellent example of how a focus on expanding access to better, faster, and cheaper tools can energize individual researchers and revolutionize a discipline, and BRAIN will follow that pattern.

8. **What are your general thoughts about "Big Science" initiatives, like the BRAIN Initiative? Specifically, do you think it will help everyone, including individual investigators that are working in fields not directly related to the aims of the BRAIN Initiative, or will it only help a select few scientists who have been advocating for this initiative?**

As mentioned, the goal of the BRAIN Initiative is to develop new tools and technologies that enable researchers to expand their understanding of brain function. Focusing on tool and technology development allows many more researchers to have access to tools that they might otherwise not be able to develop on their own. For example, optogenetics tools, which enable researchers to control nerve cell activity by light pulses, were pioneered about 5 years ago. Already perhaps as many as a thousand researchers worldwide have adopted optogenetics to study specific problems, and the number of investigator initiated NIH grants using optogenetics methods has climbed into the hundreds. The BRAIN Initiative is interacting extensively with the scientific community to determine what directions in tool and resource development will provide maximum benefit to the broader scientific community.

9. **Dr. Landis, you mention how the BRAIN Initiative can take lessons from the successful Human Genome Project, including the importance of sharing data widely. What policies, including data management and access policies, need to be put in place to make sure that data generated from the BRAIN Initiative can be shared across disciplines and the private sector?**

Data sharing, as appropriate, will be essential for the BRAIN Initiative, and NIH will implement management polices to ensure access not only to data but also to analytic tools, just as NIH has done throughout the Human Genome Project. The BRAIN Initiative is still in an early planning stage, and the specific policies that we implement will depend on the types of data researchers collect under the Initiative. The NIH BRAIN working group held a meeting with the scientific community in July on data-related issues, and NIH will develop data sharing policies and resources to enable access as appropriate to the Initiative as plans go forward. NIH will examine the successful data sharing practices from the Connectome Project, the Human Genome Project, and other projects as appropriate. For example, the centralized Human Genome Project sequencing centers made data available on daily basis, but NIH policies allow individual genetics investigators a set period of time before sharing some types of genetic data that they generate in their individual laboratories to allow them to publish on their own research. Data sharing has also been an important component of the Human Connectome Project, with datasets being made freely available to the scientific community.

67

Responses by Dr. Marcus Raichle

Questions for the Record
House Committee on Science, Subcommittee on Research and Technology
Hearing on "Frontiers of Human Brain Research"
July 31, 2013
Prof. Marcus Raichle

1. In the area of brain research, what assumptions from the past are we holding on to when looking at future research? Is our creativity being held captive by traditional notions of how the brain works?

Brain researchers are a very diverse group bringing to the task of understanding how the brain works very different perspectives. Traditionally, for the basic neuroscientist, the approach has been to understand how individual neurons work. More specifically how and when they create action potentials or 'spikes'. This work has generally been confined to work in laboratory animals where spikes from small numbers of cells can be recorded. We have learned much from this work about individual cells but less about the integrated action of the whole brain and, particularly, its dysfunction in disease.

At the other extreme we have clinical neuroscientists (neurologists, neurosurgeons and psychiatrists) who examine brain function from a clinical perspective often relying on the presence of disease in particular parts of the brain to infer function (e.g., patients with stroke who exhibit a variety of specific symptom and signs depending upon the location of the stroke) or the effect of drugs with specific actions as in the case diseases like Parkinson's disease or psychiatric illnesses like depression and schizophrenia.

And, finally, we have behavioral scientists like psychologists who examine the details of human behavior and infer from their observations how the brain works.

The future as I see it involves the integration of all of the above approaches. The prospect of this possibility emerged dramatically in the 1970s with the introduction of X-ray computed tomography or CT as it is now known. Almost overnight the practice of medicine was changed forever. CT was followed in rapid succession by positron emission tomography or PET and then magnetic resonance imaging or MRI, both offering the possibility of examining the chemistry, metabolism and function of the human brain in health and disease, a promise that is being amply fulfilled.

The challenge is achieving the integration across levels of analysis from humans to single cells and everything in between. It is, of course, true that at each of these levels new tools and significant modification of existing tools will continually emerge. But that should not be touted as re-establishing the primacy of any given level of investigation. Rather it should be viewed as strengthening the overall enterprise. However, convincing advocates of a particular approach is often difficult if not impossible. This was recognized by the James S. McDonnell Foundation and the Pew Charitable Trusts when they initiated their landmark program in cognitive neuroscience, a merger of cognitive psychology and human neuroscience-based imaging. The idea was to train a new generation of neuroscientists who understood brain science form a new perspective. To achieve this goal, mentorship of trainees had to be assigned to two individuals working at different levels of analysis, the thought being that those already committed (i.e., senior investigators) would find it difficult to achieve this integration

68

personally. The results of this initiative have been remarkable to say the least and should serve as a model of how we might approach the training of new neuroscientists. It is, I believe, the only way to avoid being captive to traditional notions of how the brain works.

2. Do private foundations have an easier time funding high-risk, high reward research in the area of brain science? Is it fair to say that the NIH funding mechanism is biased towards picking grants that are 'safer' in its aims? Please explain.

Private foundations, a unique American tradition, have brought enormous added value to the research enterprise of this country. While their resources and their missions vary tremendously they have sought to identify areas of need and unrecognized potential and provided funding to help establish the importance of these areas. But as has been said by others, the capacity to go beyond this level of support is limited. They are a catalyst but providing sustained support which is the bedrock of science is beyond their capacity. This is where NIH funding for the biomedical sciences has been critical by creating and supporting the most successful research enterprise the world has even known. The value to Americans, and more broadly, the citizens of the world, has been enormous.

It is indeed true that the traditional NIH funding mechanism (i.e., the investigator-initiated grant or R01) relies heavily on preliminary data. By that I mean, data contained in the grant application that indicates to some level of certainty that the research proposed is feasible and that the investigators have the capacity to perform that research. In this day of limited research funds from the NIH it is only prudent that research funded is likely to be successful.

The NIH is, in my estimation, very much aware of the fact that young investigators must have the opportunity to 'prove' that they have the potential to be creative and productive investigators. They provide, therefore, funds for training and 'starter' grants to help these young investigators amass the necessary preliminary data. This is a critically important funding mechanism that must be supported and sustained if the United States is to maintain its preeminent position in the biomedical sciences. Private foundations have played a minor, albeit important, role in this.

3, Are you aware, NIH spends more than $5.5 billion in the area of neuroscience research. How would the BRAIN initiative produce substantive change in disease outcome and better public health for the nation, compared to these existing efforts?

Yes, I am aware that the NIH continues to generously support neuroscience research despite the financial constraints under which it is currently operating. In evaluating this level of support it must always be kept in mind the growing financial impact of neurological disease as the number of senior citizens continues to increase. For diseases such as Alzheimer's disease and other disabling conditions this impact will be enormous. Our support for research should always be viewed in the context of the magnitude of the problem.

It is my understanding that the BRAIN initiative will add about 2% to our investment in neuroscience research at the NIH. As initially conceived, it focused on the development of new tools for the measurement of brain function. In so far as this work can inform us about the work we are currently doing it will have a positive impact. For example, imaging of brain function in humans in health and disease is dependent on our understanding of the nature of the imaging signals coming from PET and MRI. While tremendous progress has been made in this regard we have much more to learn. Much of the needed information will come from our detailed understanding of how brain cells, not just the many different types of neurons but also the many other cells types that play critical roles in brain function, are operating together. The value of the proposed work will, I believe, be directly related to the manner in which these new data are integrated into a larger sense of our goals. As I pointed out in #1 (above) an integrated understanding of scope and goals of neuroscience must be fully understood. Narrowly focused on 'my tool' is inadequate.

4. In the brain science community today, do you agree that there is a consensus about what should be mapped? What about general consensus on the best approach to mapping the brain and on which approaches should be given the highest priority? Is the BRAIN initiative calling for a lot of effort and spending based on an outmoded paradigm?

No, and that is part of the problem. There tends to be a somewhat Balkanized view depending on the level and orientation of the scientists queried. The people working at the cellular and molecular level see answers to pressing questions residing in their domain. Neurophysiologists see electricity in all of its manifestations as the defining characteristic of the brain and until fully understand progress will be limited. Cognitive neuroscientists and clinical neuroscientists see human brain imaging as the answer. And, finally, behavioral scientists often complain that too much emphasis is placed on brain science.

From my perspective the best way forward is one of integration across levels of analysis. This is no easy task and requires dialogues among the experts at all levels. Scientists need to understand the thrill of seeing their work in a larger context that is more than just hand waving about its relevance to this or that disease. And, finally, the human brain has joined the agenda of serious neuroscience research and because human brain function in health and disease is a goal for all to embrace it must be adequately supported.

5. The creative drive and engine for American Science is the individual investigator. This is one of the reasons we have a peer review process at agencies like NSF and NIH. Why then do we need the BRAIN initiative? If we have faith in the peer review system, won't these projects get funded anyway?

Individual initiative and creativity are critical components of any research project. But the questions being addressed these days in neuroscience and elsewhere often require the collective interaction of individuals with differing backgrounds and talents. This has been realized for many years at the NIH in the form of Program Projects where a general theme is pursued by groups of individuals working together, comparing ideas, addressing challenges and doing experiments in a broader context than might normally exist in a single investigator initiated research project. Some very important research lends itself to this approach other research does not.

A great example of a collaborative endeavor is the NIH sponsored Human Connectome Project. This project endeavors to create a normative database of human brain imaging involving anatomy and function that includes a variety of information including genetics. The project involves the best investigators available in the diverse areas of expertise needed for this enormous undertaking. While the majority of individuals involved come from the United States other countries, which unique expertise resides, are included as well.

While I have not followed in detail the programmatic development of the BRAIN initiative at the NIH, it is my understanding that focus of the work will be on new techniques and model systems of brain function. Much work on some of the techniques proposed is well underway. Other aspects such as nanoparticle sensors of neuronal function are still very much in the developmental phase. From my perspective, it is important to understand the goals of the initiative in the larger context of neuroscience research and its ultimate goal of understanding human brain function and behavior in health and disease.

6. What are your general thoughts about "Big Science" initiatives, like the BRAIN initiative? Specifically, do you think it will help everyone, including individual investigators that are working fields not directly related to the aims of the BRAIN initiative, or will it only help a select few scientists who have been advocating for this initiative?

The Human Connectome Project is a great example of "Big Science" that will benefit many investigators. This project was specifically designed to place at the disposal of all investigators interested in human brain function a body of normal data that would be difficult if not impossible for many to collect. This follows a pattern of such projects in areas like Alzheimer's disease where the data from the Alzheimer's Disease Neuroimaging Initiative (ADNI) and the Dominantly Inherited Alzheimer's Network (DIAN) provide unique data (e.g., imaging of brain anatomy, function and metabolism; genetics; behavior; and biomarkers). The required sharing these data with all interested and qualified investigators is an enormously powerful means of obtaining the maximum return on the research dollars invested. Given the fact that these "Big Science" research initiatives were funded by our tax dollars it is a means of maximizing the return on our investment.

The BRAIN initiative, as a very high profile undertaking, must strive to define the benefits to be derived from an integrated approach to methods development. There is no question that methods are critical to any research enterprise. A primary example is imaging itself (i.e., CT, PET, and MRI). It has transformed human neuroscience. Will these new methods and improvements in existing methods (e.g., MRI) really benefit from this focused approach? I am confident that individuals now concerned with the implementation of this initiative from the Director on down are very cognizant of the challenge. The other benefit to accrue from the BRAIN initiative is to focus the publics' attention on both the importance and excitement of brain research. As many have said, the human brain is the last great frontier of science and one that we ignore at our peril.

Responses by Dr. Gene Robinson

**QUESTIONS FOR THE RECORD
THE HONORABLE ELIZABETH H. ESTY (D-CT)
U.S. House Committee on Science, Space, and Technology
Subcommittee on Research**

The Frontiers of Human Brain Research
Wednesday, July 31, 2013
Questions for Prof. Gene Robinson

Thank you, Chairman Bucshon and Ranking Member Lipinski for holding this important hearing on brain science research. As we have heard this morning, brain science research is imperative as we work to provide medical care to our aging population. In my district alone we have more than 99,000 seniors, and research like this is necessary to understand and combat neurodegenerative diseases like Parkinson's and Alzheimer's. And it sounds like, from what we have heard so far today, that scientists have made great strides in battling these diseases.

Dr. Robinson, you spoke earlier about the role of supercomputers and data collection in brain science research. In your opinion, what is the role of big data and supercomputing as we continue to explore the brain and dig deeper into brain science research?

The human brain is an assemblage of almost 100 billion interconnected brain cells (neurons), whose individual activities are integrated to produce our thoughts and behaviors. To understand how this works, we need huge data sets that describe how different types of neurons function in different situations. We also need powerful supercomputers that can synthesize and make sense of those data sets to give us new insights into how the activities of individual neurons give rise to a fully functioning brain.

The brain is an incredibly complex system, one that is dynamic on multiple scales—from the many connections (synapses) that each neuron makes with other neurons, to the ensembles of neurons that act together to form circuits, to the activity of the whole brain. Our goal as neuroscientists is to understand this system well enough to predict how the brain will respond in any situation, to any type of input. This can be accomplished by collecting data about how the brain works in different situations, and then synthesizing those data into a more general conceptual model of how brain activity on the level of single neurons sums to produce a whole brain response. However, it will take a large body of precise and accurate data to successfully build this model. We have some of the tools we will need to produce it—tools that let us observe or manipulate the brain or its component parts—while others have yet to be developed. To synthesize so much data from different sources and functional levels in the brain, we will also need to push the current limits of computing power. Blue Waters, the University of Illinois's NSF-funded supercomputer, is an example of the type of instrument we will need to model how brains work to produce behavior.

How has the collection of big data helped to make interdisciplinary research more accessible?

A good example of how big data has changed the way scientists work together comes from genomics. When scientists first determined the sequence of the human genome, the complete set of human genetic material, they produced the largest biological data sets that had ever existed. These data sets not only invited an interdisciplinary approach, they demanded one. A genome is both a blueprint for the functioning of a living cell, and a historical record of the evolution of a species and an individual organism; there is more raw data in it than one research group could ever hope to make sense of. To explore all the meaning hidden inside the genome requires collaboration among researchers who represent biological study at every level, from molecules to whole organisms to ecosystems. Furthermore, the statistical methods, computational algorithms, and software that were once developed to work with individual genes could not handle the amount of data in a complete genome. Biologists needed the partnership of mathematicians, computer scientists, statisticians and others to help develop the requisite new tools. As DNA sequencing becomes ever faster and cheaper, and scientists seek new insights by determining and comparing the genome sequence of a broadening set of species, they will continue to need new and better technologies to handle the data. The CompGen Initiative, which was recently awarded funding from NSF, is one example of an interdisciplinary effort to address this need. In this initiative, University of Illinois biologists, computer scientists, statisticians, and engineers will be working together to create an instrument with hardware and software designed to handle the challenge the huge data sets that are becoming commonplace in genomics. Big data sets in any field create challenges that are best met with this type of interdisciplinary approach, and the success of data sharing and collaboration in genomics will continue to inspire more interdisciplinary work in other fields, including neuroscience.

Appendix II

ADDITIONAL MATERIAL FOR THE RECORD

SUBMITTED STATEMENT BY CHAIRMAN LAMAR S. SMITH

Thank you Chairman Bucshon for holding this hearing.

The brain is a fascinating subject, and one of the unknown frontiers of medical science. We all have a brain, but we barely understand how it works.

But through the process of science, we have begun to understand what questions to ask, what tools we need and the complexities that underlie the trillions of connections between neurons.

Developments in basic scientific research, such as those contributed by Prof. Marcus Raichle, have provided deep insight into how the brain is organized.

As the witnesses will discuss today, brain science is inter-disciplinary in nature. Advances from applied mathematics, physics, chemistry, computer science and engineering help provide both a conceptual understanding and experimental tools.

In my view, this is where the National Science Foundation (NSF) can play an important role towards understanding the basic science behind Alzheimer's, Parkinson's, autism, stroke, dementia, traumatic brain injury, epilepsy and many other debilitating neurological disorders.

I believe the NSF should support inter-disciplinary research in this area because the results of this research will have clear and direct benefits to the American people.

The results of this research could be the foundation of new technologies that help wounded warriors walk again and also improve the quality of life for many injured Americans.

For example, near my district in San Antonio, the Department of Orthopedics & Rehabilitation at Brooke Army Medical Center provides state of the art orthopedic and rehabilitative care to active duty soldiers of all services. I have met many of these wounded veterans who deserve a better life.

My district is also home to several brain rehabilitation centers, including the Texas NeuroRehab Center and Reeves Rehabilitation Center. These centers treat thousands of patients who look forward to leading independent and productive lives.

Research the NSF funds in robotics, statistics, fast algorithms and computation can be used by medical doctors to help patients perform day to day tasks.

This past April, the Administration announced the Brain Research through Advancing Innovative Neurotechnologies Initiative, otherwise known as the BRAIN initiative. While I do not think many would disagree with the goals of this initiative, I am concerned that this is solely a repackaging of existing initiatives.

Any federal initiative should include stated hypotheses along with clear steps towards implementation.

I hope this hearing serves as an opportunity to work together and look for a bipartisan solution to funding inter-disciplinary brain-related research.Thank you Mr. Chairman for holding this hearing, and I look forward to the witnesses' testimony and questions. And I yield back.